I0123869

Gendering the Field

Towards Sustainable Livelihoods for Mining Communities

Asia-Pacific Environment Monograph 6

Gendering the Field

Towards Sustainable Livelihoods for Mining Communities

Edited by Kuntala Lahiri-Dutt

ANU

THE AUSTRALIAN NATIONAL UNIVERSITY

E PRESS

ANU E PRESS

Published by ANU E Press
The Australian National University
Canberra ACT 0200, Australia
Email: anuepress@anu.edu.au
This title is also available online at: http://epress.anu.edu.au/gendering_field_citation

---

National Library of Australia
Cataloguing-in-Publication entry

Author: Lahiri-Dutt, Kuntala.

Title: Gendering the field : towards sustainable
                 livelihoods for: mining in mining communities /
                 edited by Kuntala Lahiri-Dutt.

ISBN: 9781921862168 (pbk.) 9781921862175 (eBook)

Series: Asia-pacific environment monographs ; 6.

Notes: Includes bibliographical references.

Subjects: Mineral industries--Participation, Female.
                 Women in sustainable development.
                 Mineral industries--Employees.
                 Households--Economic aspects.

Dewey Number: 338.2082

All rights reserved. No part of this publication may be reproduced, stored in a retrieval system or transmitted in any form or by any means, electronic, mechanical, photocopying or otherwise, without the prior permission of the publisher.

Cover design and layout by ANU E Press

Cover image: KPC employee for 16 years, Rina is a woman of Dayak ethnicity from Central Kalimantan. Photograph taken in 2008 by Kuntala Lahiri-Dutt.

This edition © 2011 ANU E Press

Contents

List of Figures

List of Tables

List of Appendices

Foreword

Though women all over the world toil as miners, and have done for centuries, mining is seen as a quintessentially masculine endeavour. *Gendering the Field* puts a definitive stop to the gender-blindness of such a view. We have Kuntala Lahiri-Dutt, the editor and a contributor to this book, to thank for this corrective move.

For many years Lahiri-Dutt has championed the cause of women in mining. She has challenged the preoccupation of minority world feminist scholarship with women living in mining communities and argued for more analysis of women working at the actual mine-site. Her interest in women in mining grew out of important research into artisanal small-scale mining on the coal fields of West Bengal. Here women to this day work with men scavenging for coal, wheeling it to market in improvised vehicles, living alongside their pits and conducting family life in the interstices of mine work. Such a world seems far from the large-scale mines that occupy a privileged place in the global imaginary. In our resource hungry world we are familiar with the huge excavators operated by well-paid men gouging out minerals, moving mountains and transforming landscapes in a matter of years. Or the construction of ever deeper underground mines where sophisticated computer models guide the operations of advanced machinery 'manned' by fly-in/fly-out miners working 12 hour shifts. To this technology intensive, heroic, and above all masculine landscape Lahiri-Dutt has added a panorama of poor women and men using rudimentary tools to produce vast amounts of mineral output, outside the regulatory embrace of union organisation and health and safety laws. Exposés of the extent of this artisanal mining 'industry', such as those by Lahiri-Dutt for India, have forced a long overdue shift in international mining policy and planning.

Widening our perspective so that we can see what is in front of us differently is one of the key contributions of feminist scholarship. Another is adherence to an action oriented mode of research. No matter how theoretical, empirical or pragmatic, feminist scholarship seeks to change the world in which we live, bringing greater dignity and justice to those who experience discrimination. Lahiri-Dutt's research has done this for small-scale artisanal miners. But she has not stopped there. From this situated research experience she has embarked on an impressive action-oriented engagement with the mining industry as a whole. *Gendering the Field* is a testament to her determination to leave no shaft in these complex workings unexplored. The authors she has collected together in this volume explore crucial gender issues across the board of large and small mining operations in North America, Canada, Australia, Papua New Guinea, Indonesia, Mongolia and the Democratic Republic of Congo. Chapters include perspectives of indigenous women and women in support industries such as sex work and other input suppliers. All contributors are concerned to address the community level development implications of mineral extraction.

The collection offers a refreshing engagement with contemporary gender and development theorisations and their application in the 'field' of mining. While there is a strong element of criticism of the status quo in the mining industry with respect to the treatment of women, there is also an even-handed appreciation of the attempts being made by corporations and governments to address inequalities and listen to the concerns of women, particularly indigenous women, involved in mining operations. The collection contains both healthy scepticism and generous acknowledgement of the advancements made in the name of sustainable development in mining areas.

Perhaps the most difficult issue confronting social scientists concerned about gender equality, economic justice and sustainability in the global mining industry is that of the non-renewable and thus inherently unsustainable nature of mining activity. As with many political victories, the timing for women is all wrong. Having been ousted from underground mines as the industrial revolution gathered pace in the northern hemisphere minority world, women have only relatively recently been accepted back into the mining workforce, just as the mining industry has become perhaps *the* major political hurdle to addressing the social and environmental challenges posed by climate change. And in majority world nations where mineral riches are increasingly being exploited on a large scale for international export, mining development offers women and men alike a real chance to get ahead. At the same time, there is greater realisation of the devastating consequences for proximate ecologies and global environments of such scaled-up operations. What's more, there is little evidence that the exploitation of mineral riches in poor nations leads to greater wellbeing. So, as many of this volume's authors admit, development itself is under question.

What might sustainability mean in connection with the international mining industry? This important book provides an answer to this question by outlining ways of thinking about the community livelihoods that might be supported alongside minerals development. Shifting the centre of attention from sustaining a controversial 'development' to sustaining the livelihoods of women and men in surrounding communities opens up the space to allow a range of practical initiatives. For example, procurement practices can distribute the economic benefits of mining to a wider region and to women, and negotiated agreements can safeguard the spiritual and social needs of indigenous communities living in the vicinity of mining activities. There needs to be many more strategies for transforming a non-renewable activity that destroys landscapes into renewable economic practices that support livelihoods and replenish ecologies. As women enter the mining industry in greater numbers and as their longstanding contributions in artisanal small-scale activities gain greater recognition it is timely to raise these important and hard issues of planetary survival. While there is no necessary connection between women and sustainability, it remains a commonplace observation that where women are able to survive well, their families and communities survive well too. The combined attention to gender and sustainable livelihoods of this volume signals an important turning point, not only for mining industry scholarship, but also hopefully for minerals development policy. We now know that mining companies can attend to gender equality and still perform well in the marketplace. They can also support community livelihoods and compete in the cut and thrust of a competitive industry. Perhaps the next challenge is to see how they can also become responsible and reparative environmental citizens and survive as businesses. The evidence provided in this volume allows us some hope for the future.

Professor Katherine Gibson

Centre for Citizenship and Public Policy

University of Western Sydney

December 2010

Glossary

adivasi [Hindi]	indigenous people
bidi [Hindi]	indigenous type of cigarette
ikut suami [Bahasa Indonesia]	to follow your husband
kerja begini [Bahasa Indonesia]	euphemism 'working like this'
le mariage précoce [French]	forced marriage
lokalisasi [Bahasa Indonesia]	quasi-official brothel complex
ojek [Bahasa Indonesia]	moped
wanita nakal [Bahasa Indonesia]	naughty/bad women
warung [Bahasa Indonesia]	cafe
wisma [Bahasa Indonesia]	brothel

Abbreviations

ABCD	Asset-Based Community-Driven
ALP	Ahafo Linkages Program
ATSIC	Aboriginal and Torres Strait Islander Commission
ASM	artisanal and small-scale mining
CEDAW	Convention on the Elimination of Discrimination against Women
CEP	Coal Employment Project
CDT	Community Development Toolkit
CSR	Corporate Social Responsibility
DFID	Department for International Development
DRC	Democratic Republic of Congo
DDR	Disarmament, Demobilisation, and Reintegration
EEO	Equal Employment Opportunity
GAD	gender and development
GEM	Gender Empowerment Measure
GDI	Gender-Related Development Index
GDP	gross domestic product
GRI	Global Reporting Initiative
HSEC	Health, Safety, Environment and Community
HDI	Human Development Index
ICMM	International Council on Mining and Metals

IFO	International Finance Corporation
ILO	International Labour Organisation
IUMMSW	International Union of Mine, Mill, and Smelter Workers
IWMN	International Women and Mining Network
KLC	Kimberley Land Council
KPC	PT Kaltim Prima Coal
LAST	Livelihood Asset Status Tracking
LIA	Labrador Inuit Association
MDG	Millennium Development Goal
MMG	Mineral Metals Group
MMS	Mining and Metals Sector
MMSD	Mining, Minerals and Sustainable Development Project
MRPAM	Mineral Resources and Petroleum Authority of Mongolia
NCGE	National Committee on Gender Equality
NDS	National Development Strategy
NGGL	Newmont Ghana Gold Ltd
NGO	non-government organisation
PNG	Papua New Guinea
PPP	Purchasing Power Parity
RTN	right-to-negotiate
SAESSCAM	Small-Scale Mining Technical Assistance and Training Service
SDC	Swiss Agency for Development and Cooperation
SGBV	sexual and gender-based violence
SMEs	small and medium enterprises
TIA	Tongamiut Inuit Annait
UNIDO	United Nations Industrial Development Organization
UMW	United Mine Workers
WID	Women in Development

Currency Conversion Rates

Year	USDAUD	USD/CDF	USD/IDR
2006	0.7884	n.a.	7500
2007	0.8776	n.a.	n.a.
2008	0.6983	826	n.a.

Contributors

Sara Bice is completing her Doctoral degree in Sociology at the University of Melbourne, Australia. Her thesis is examining ways to measure the social impacts of Australian mining companies' operations in developing countries. Sara has worked in international and community development, especially on women's issues with Queen Victoria Women's Centre, Brotherhood of St Laurence and Oxfam Australia. Sara holds an M.A. in Gender and International Development Studies (Melbourne) and a B.A. in Journalism and Mass Communication (UNC-Chapel Hill, USA, Hons).

Gill Burke is a mining historian and economist. She holds a PhD from London University for a thesis on the nineteenth and twentieth century Tin Mining Industry of Cornwall, UK. From 1985, Gill worked as consultant economist and technical assistance advisor in the field of mineral resources development—mainly in Southeast Asia. In this context she worked for international agencies such as UNDP, UNIDO and the ILO as well as for private sector and state mining corporations. Over time Gill became a specialist on artisanal, small-scale and illegal mining with concomitant interest in mining and gender. In 2001 she was part of the team evaluating the mining industry of Papua New Guinea for the SYSMIN Fund of the European Union; with gender and small-scale mining issues as her particular responsibility. In 1990 Gill came to Australia as a Visiting Fellow in the Research School of Pacific and Asian Studies at The Australian National University. She maintains links with ANU, especially the Resource Management in Asia-Pacific Program in the College of Asia and the Pacific. In addition she still does consultancy work for the Raw Materials Group in Sweden.

Ana Maria Esteves is a consultant in the resources sector specialising in corporate-community investment, social impact assessment and community engagement. Originally from Mozambique, Ana Maria has been a resident of Zimbabwe, South Africa, Portugal, Hong Kong and Australia. She has an MBA from Melbourne Business School and Ph.D. from the University of Melbourne. Her thesis addressed how resources companies can contribute to social development while building business value.

Kuntala Lahiri-Dutt is a Fellow at the Resource Management in Asia Pacific Program at The Australian National University. Kuntala has conducted extensive research on the gender, social and environmental issues related to mining since 1993. She has written widely on gender in both large-scale and small mining sectors as well as on developmental challenges in areas of mining operations. Her books on mining include *Social and Environmental Consequences of Coal Mining in India* (jointly edited with G. Singh and D. Laurence, 2007); *Women Miners in Developing Countries: Pit Women and Others* (jointly edited with

Martha Macintyre, 2006); and *Mining and Urbanization in the Raniganj Coalbelt* (2000). Her research papers on topics related to mining have been published, among others, in *Feminist Review, Natural Resources Forum, Economic and Political Weekly* and *South Asian Survey*. Besides mining, Kuntala has also had long-term research interests in water and gender, and community participation in water management.

Martha Macintyre recently retired as Senior Lecturer in Medical Anthropology at the Centre for Health and Society, University of Melbourne. She has held research positions at Monash University and The Austalian National University and taught Anthropology and Women's Studies at La Trobe University. In 2005 she was elected President of the Australian Anthropological Society. Her research interests include historical ethnography of the Pacific region; anthropology of gender; immigrant communities in Australia; development studies; and medical anthropology. Her ethnographic research has concentrated on Milne Bay Province and New Ireland Province in Papua New Guinea, where she has studied social and economic change associated with colonisation, missionisation and economic development. She has also worked as a consultant, evaluating the social impact of mining projects on Lihir and Misima in Papua New Guinea.

Petra Mahy is completing her doctoral studies in the Resource Management in the Asia-Pacific Program at The Australian National University. Since 2006 she has conducted her field research in East Kalimantan, Indonesia, on gender and community development in a coal-mining area, focusing on issues of decentralisation and corporate social responsibility. She has undergraduate qualifications in Arts, Asian Studies (Indonesian) and Law.

Laurie Mercier is Professor of History in the College of Liberal Arts at the Washington State University and teaches the history of the United States, the American West, the Pacific Northwest, immigration and migration, and American labour. She is the former associate director of the Center for Columbia River History, a former president of the Oral History Association, and co-director of the Columbia River Basin Ethnic History Archive project. Her research and publications focus on workers and labour, identity, gender, region and community. Her recent publications include, *Anaconda: Labor, Community, and Culture in Montana's Smelter City* (2001); 'Reworking Race, Class, and Gender into Pacific Northwest History' (*Frontiers: A Journal of Women's Studies*, 2001); and 'Instead of Fighting the Common Enemy: Mine Mill and the Steelworkers Unions in Cold War Montana' (*Labor History*, 1999).

Ciaran O'Faircheallaigh is Professor and Head of the Department of Politics and Public Policy at Griffith University, Brisbane. He has previously held research and teaching positions at The Australian National University, the University of Papua New Guinea and Queens University, Ontario. During

1998-2000 he was Policy Advisor to the Queensland Indigenous Working Group, Queensland's peak Indigenous organisation for native title and cultural heritage matters. His research focuses on the conduct of policy and program evaluation, particularly in areas where policy goals are highly contested; on the management and impact of public programs and policies designed to meet the needs of Australia's Aborigines and Islanders; and on negotiations between commercial interests and Indigenous peoples in Australia and Canada. He has advised Aboriginal organisations in Australia on negotiating mining agreements for over a decade and was Senior Consultant, Major Projects, to the Cape York Land Council from 1995 to 2001. He has published in a wide range of journals including *World Development, Society and Natural Resource, ResourcesPolicy* and *Energy Policy.* His most recent book, *Earth Matters: Indigenous Peoples, The Extractive Industries and Corporate Social Responsibility,* (2008) was co-edited with Saleem Ali.

Joni Parmenter is a research officer with the Centre for Social Responsibility in Mining, University of Queensland where she undertakes research into the social dimensions of mining. She has particular research interest in the impacts of mineral development and mine closure on Indigenous communities and how these communities can achieve more sustainable outcomes from mineral development. Joni is also interested in gender issues around mining and communities, as well as general diversity and gender issues within the mining workforce. She holds a B.A. (Hons) in Anthropology from the University of Queensland.

Rachel Perks is the Conflict Program Manager within the Extractive Industries for the international non-government organisation, Pact in the Democratic Republic of Congo (DRC). Rachel holds an Honours degree in Peace and Conflict Studies from the University of Toronto and is currently undertaking her doctoral research in the School of Agriculture, Policy and Development at the University of Reading. She has worked for the last eight years in Sudan, northern Uganda, Kenya and the DRC and has extensive experience working on peace processes, reconciliation and post-conflict transitions in African countries. Conflict in the DRC has traditionally been, and continues to be, in large part over natural resources. Pact manages a public-private partnership with four internationally-listed companies, as well as the US Agency for International Development (USAID) and the DRC Government for sustainable and responsible development of communities within the context of the mining sector.

Bolormaa Purejav is an economist and holds an M.B.A. degree from Griffith University, Australia. She has worked for numerous government agencies and international organisations within Mongolia, including the UNDP and Swiss Agency for Development and Cooperation. She has coordinated three gender research projects for the Mongolian government, including Capacity Building for Gender Sensitive Budgeting and contributed to the 2007 UNDP Human development report. She is currently the Planning, Monitoring and Evaluation Officer for the Mongolian Sustainable Artisanal Mining Project, responsible for mainstreaming a community development approach throughout all project components as well working with artisanal miners in the field.

1. Introduction: Gendering the Masculine Field of Mining for Sustainable Community Livelihoods

Kuntala Lahiri-Dutt

This volume presents a selection of papers that were presented at an international workshop on 'Mining, Gender and Sustainable Livelihoods', organised to disseminate the results of an 'action research' project.[1] The project endeavoured to integrate a gender outlook in one major mining company's community development initiatives, and strengthen interdisciplinary approaches in examining the interface between gender, mining and sustainable livelihoods (EC 2007). Held in late 2008 in the Resource Management in Asia-Pacific Program (RMAP) of The Australian National University (ANU) in Canberra, the workshop was, for a number of reasons, an important international event in the field of social and community issues surrounding the extractive industries. It represented a confluence of several streams of thought and disciplinary approaches to gender and mining. The significance of this confluence lies in its holistic approach, through a conflation of community and gender interests, to the broad field of mining—without separating large and formal mining from informal, artisanal and small-scale mining (ASM) practices. Academics and other researchers, activists, civil society organisations and policy-makers who participated in the workshop joined hands to address the increasingly important question of how to engender the field of mining. Above all, the workshop brought together, on the same platform, the collective wisdom of many of the key people actively working on mining and gender in a wide variety of contexts, with different lenses and from different perspectives. The chapters in this volume[2] highlight the key issues and implications of integrating gender to foster sustainable livelihoods in both large-scale and informal, artisanal and small-scale mining in different parts of the world.

---

1   The 'Mining, Gender and Sustainable Livelihoods' workshop was the outcome of an action research project funded in 2006 by the Australian Research Council. This Linkage Project was a collaboration between The Australian National University (ANU) and PT Kaltim Prima Coal (KPC). Project outputs included in-company gender surveys, gender needs assessments, gender impact assessments, social impact studies and gender training materials. As a Chief Investigator for the project, I would like to acknowledge the industry leadership that KPC provided in recognising and responding to gender concerns and opportunities in their organisation and community activities. In addition to the Australian Research Council, we received funding from the World Bank and AusAID to support the travel of international participants; I thank both these agencies for this assistance.
2   This volume represents only a selection of papers that were presented in the workshop. The manuscript has gone through the usual peer review process, and I thank the reviewers for their constructive suggestions and comments.

A significant body of scholarly and practitioner research has now emerged on this niche area of interest, namely gender and the extractive industries. Examples and case studies from all around the world have drawn attention to the need to make visible the roles of women in large-scale mining operations, and to encourage government and industry to adopt inclusive community development processes to ensure that mining-led development can benefit women and men equally. Another expanding body of research and action has drawn attention to the contribution and roles of both women and men in informal mining, more commonly known as ASM—a diverse range of mining practices that sustains the livelihoods of millions of people in the mineral-rich tracts of poorer countries around the world. This evidence has far-reaching implications for future research, activism and industry practice with regard to sustainability and policy-making by national and international bodies, development agencies and mining companies. It is of great importance that the researchers are able to communicate their research and experience, think critically and reflect on the initiatives that have been undertaken in different contexts by different actors, and ask whether the ideas developed in one corporate, political or cultural context can or should be applied to others.

One of the questions addressed by the chapters in this volume is what scholarly and practitioner research—using historical, feminist or contemporary participatory approaches—can contribute to the growing body of knowledge on sustaining community livelihoods in areas of mining and other extractive practices. This remains a highly contested terrain, and the papers in this volume expand the vista of contemporary theorisations situated upon it. They add depth to international-level policies and practices in the industry that have begun to initiate some corrective measures in this regard. The papers in this volume contribute to exploding the myth of the gender-equitable and homogeneous 'community' that other natural resource sectors discarded decades ago, but that has persisted in the extractive industries. The volume considers effective ways to integrate gender into these industries' development project to contribute to meeting Millennium Development Goal 3, to promote gender equity and to empower women, in areas of mining work. What is the relationship between gender equity and sustainable development, and is it universal in nature? A critical issue, raised by Macintyre in this volume, is what our approach should be in challenging the gender-blind, male-centred conceptualisations of development, promulgated by the extractive industries. In responding to the marginalisation of women in these industries' community practices, should one adopt a 'feminising development' approach? Indeed, in our joint work published earlier, we took up such a women-centred approach while suggesting a 'human rights of women' case (see Lahiri-Dutt and Macintyre 2006). However, experience has shown us that such an approach, best expressed in the genre of 'Women in Development' (WID) philosophy, has the potential of diluting the feminist agenda through its uncritical stance on development.

The adoption of WID by addressing so-called 'women's issues' (think, for example, about community health projects stressing contraception and fertility) allows the mining and other extractive industries to utilise instrumentalist justifications for 'incorporating' women, and leads to a shift away from equity to efficiency approaches in development projects. As we now know, such 'add women and stir' projects primarily address women's practical, rather than strategic, gender needs and interests through women-only projects and organisations. Since the 1980s, a growing consensus among feminists has highlighted the need to shift the focus from 'women' to 'gender'. McIlwaine and Datta (2003: 370–1) feel that a gender and development (GAD) discourse offers a more radical agenda that can address the bases of inequalities between women and men, and redistribute the power inherent in gender relations. Indeed, in integrating gender concerns in the extractive industries, the adoption of a GAD offers significant advances on 'women only' approaches on two fronts. First, it is robust enough to critique the development process itself by highlighting how capitalist mining and industrial development adversely affect both the productive and reproductive lives of women. Gender is also conceptualised as a dynamic social construct, reflected in a greater appreciation of diversity in GAD, and can lead to shifts towards empowerment and participatory processes. Above all, as the 'natural home' for questions of race, GAD creates a space for understanding and accommodating 'difference' within its community of practitioners, allowing the creation of new and different identities for women from less developed countries. These women no longer need be uniformly victimised, poor and uneducated—the opposite of modern, liberated and educated 'Western' women.[3] In GAD, the appreciation of diversity among women, especially around race and ethnicity, has a transformative politics at its heart that is also materially engaged. This engagement is one way to reduce the gaps between research and activism. The action research project that led to these deliberations at the workshop provides an example of integrating gender in community development projects. The project itself represented a politically engaged and socially relevant form of research; a kind of research that not only used the participants to seek data but also involved and sensitised them, and in the process, created new gender identities.

---

3   White (2006: 58) comments sharply on this stark difference in visibility of race in contemporary gender studies, described by Moore as 'The whiteness of faces and Britishness of passports' in development agencies such as Department for International Development (DFID): 'while gender is marked in bright colours, race is at best a shadowy presence in development discourse.'

If, indeed, there is a case for gender mainstreaming in the extractive industries, then one may ask: what is to be mainstreamed and where? As Macintyre points out gender is often nowhere, when it is believed to be everywhere; that is, gender is mainstreamed in name but gender equality is not specifically institutionalised by embedding gender-sensitive practices and norms in policy structures, processes, and environments. The diverse nature and complexity of issues raise serious questions for the extractive industries. Are the cases for gender mainstreaming different for large and small-scale mining or are there significant overlaps? Can the adoption of a broader sustainable livelihoods approach, by incorporating gender-equitable, socially-just, pro-community, and equity-oriented development for host communities, be useful for the mining industry? What roles might the different actors—governments, international policy bodies, and the industry—play in helping to mainstream gender in the mining sector? What changes in policy or practice can be expected to arise from the growing interest in gender equity and mainstreaming? To continue with the momentum, what lessons have been learned, what further steps need to be taken, and by whom? To begin answering these questions, one must first take a brief look at what has been done so far by international policy-making bodies.

The year 2005 was a notable one in regard to how international policy bodies think about these complex and contested issues around mining. The International Council on Mining and Metals (ICMM), the global body for the extractive industries, published a Community Development Toolkit (CDT) for mining companies to use in their outreach work. For those working on the complex interface between mining and gender from a development perspective, the CDT was disappointing because of its complete neglect of gender analysis. Clearly, 'the community' envisioned in the document was an imaginary, homogeneous entity, in which the gender of the actors did not make any difference. One cannot be sure if this was indeed a deliberate omission or just an oversight. Either way, this manner of thinking reflects the assumptions of many in the industry that the interests of some people within a community can be conflated with the interests of all. Such omissions from important international bodies have the potential to reinforce the masculinism that the large-scale mining industry has been notorious for. More importantly, the CDT appears to be based on, and to perpetuate, the naïve assumption that if the benefits of mining extend to men they will automatically trickle down to women. Feminists discarded this assumption with contempt long ago, and have produced a rich body of evidence to contradict it. Unless the extractive industries build policies on and in dialogue with contemporary feminists, the path to sustainability will remain elusive.

Researchers are aware that the notional 'community', when used as a reductionist collective noun, hides many critical differences and divisions within it and is hence misleading (see Gujit and Shah 1999 for example). As Robert Chambers (2008) has pointed out, there are biases within every community that need to be recognised and offset, and attitudes and behaviours which are dominating and discriminatory are more common among the powerful. Most often, the more powerful individuals within a community are men. Maguire (1996: 29–30) observed that a gender-blind approach is common even in more participatory approaches to 'community' development: 'Gender was hidden in seemingly inclusive terms: "the people", "the oppressed", "the campesinos", or simply "the community". It was only when comparing … projects that it became clear that "the community" was all too often the male community.' By ignoring these realities, the CDT turned itself into a blunt tool, without cutting edge impact.

A significant amount of scholarly work had been published on gender, development and mining prior to the publication of the CDT. While there has been research to make visible women's past and present roles and contributions to mining—within both the industry and mining communities—there has also been a steady stream of policy-related work that has highlighted why and how development interventions work better when the benefits reach both women and men. In recent years, human rights organisations and activists have consistently argued for a rights-based approach to development and have demanded that women be included as equal partners in participatory approaches to resource management. The research project that gave rise to the 'Mining, Gender and Sustainable Livelihoods' workshop has also contributed to this growing body of knowledge and examples of practice.

The ground covered in this volume is complex and some of the terms used remain highly contested. First, while the book does not directly deal with the largely unresolved and hostile relationship between mining and development, it does favour a move towards the concept of 'sustainable livelihoods'[4] and away from the 'sustainable development' framework more commonly adopted by the mining industry. The latter framework, while useful in raising awareness of, and developing concrete steps to, care for the environment, has neglected crucial areas of concern to the extractive industries. These include unequal power relations and access to income within the community, assets that are valued

---

4   In this book, we invoke the best-known definition of sustainability or sustainable development put forth by the World Commission on Environment and Development, defining sustainability as 'forms of progress that meet the needs of the present without compromising the ability of future generations to meet their needs'. The connotation of the term, 'livelihood' is more people-centred, smaller in scale, broader as a concept and reaches beyond the economic wellbeing to the overall process of securing a survival as both food and cash by the individuals or families. Valdivia and Gilles (2001: 7) see livelihood strategies as a portfolio of activities and the social relations by which families secure or improve their wellbeing or cope with crises. To Blaikie et al. (1994: 9), the word means the command an individual, family or other social group has over an income or bundles of resources that can be used or exchanged to satisfy its needs. This may involve information, cultural knowledge, social networks, legal rights as well as tools, land or other physical resources. This definition clearly indicates that livelihood is much more than just the financial resource.

by the community, activities that sustain the community and the community's entitlements, which are linked to legal and customary rights. The volume aims not only to gender-sensitise the mining sector, but to broaden the definition of sustainability that is currently in use by extractive industries as well as development agencies engaged with mining communities or operating in mineral resource-dependent countries and regions. A broad definition of sustainability would not ignore but include, within the broad definition of 'extractive industries', ASM practices. ASM forms an important source of livelihoods in many mineral-rich tracts in developing countries, in which women continue to participate in large numbers in contrast to large-scale mining operations where they still largely remain less visible.

Second, the implication or functional meaning of the term 'gender mainstreaming' has been and remains widely debated. Gender mainstreaming can be understood as the institutionalisation of gender equality that is achieved by 'embedding gender-sensitive practices and norms in the structures, processes and environment of public policy' (Daly 2005: 435). Gender mainstreaming can also be seen as an end, rather than a means for achieving gender equality (Eveline and Bacchi 2005). For the mining sector, the definition of gender mainstreaming that most readily applies is that of an organisational strategy to be applied internally and externally as a means of bringing a gender perspective to all aspects of an organisation's policies and activities, thus building gender capacity and accountability (Reeves and Baden 2000; Walby 2005). Such an approach to gender mainstreaming would involve bringing a gender perspective and integration of gender analysis into all stages of the design, implementation and evaluation of all projects, policies and programs (UNDP 2007). Thus, we can see gender mainstreaming applying both within the industry and in the community development work of the industry, in making both women and men visible in their roles and contributions, in giving voice to women as well as men, in ensuring women's participation in community consultations and in extending the benefits of mining to them.

The range of definitions of gender mainstreaming raises the question of how such a contested concept can be crystallised within a corporate body. Moving gender from the margin of vision to the centre is a major challenge and leads to strong resistance from within organisations. On the one hand, many corporate bodies have mandates that do not match the ideology of gender equity and have bureaucratic procedures that border on inflexibility; on the other hand, gender advocates must rely on bureaucrats who are either indifferent or hostile to perceived political interference across professional boundaries into their personal lives (Razavi 1997: 1111). The international political economy of gender mainstreaming is also complex. It sounds like nearly an impossible task to bring about gender equity within corporate bodies so ensconced within the neo-liberal market-led growth paradigm. No matter what the various industry brochures and toolkits propound, philosophically the industry is inherently opposed to all kinds of interventionism because of its faith in markets as self-optimising mechanisms. Framing gender equity within mainstream neo-liberal policies gave rise to the 'gender efficiency' discourse—demonstrating positive spin-offs in terms of financial effectiveness—which has been widely critiqued (see Razavi 2009). More lately, gender mainstreaming advocates have focused on transforming the mainstream through ingenious ways, often building alliances with men.

To put my view simply, principles of gender equality can be applied vertically and horizontally across the spectrum of business activities within a mining organisation. The vertical approach would involve the creation of a separate unit or department to deal with gender equity issues, while the horizontal approach would require the application of a 'gender lens' to every aspect of an organisation's work. This should not be restricted purely to community relations departments. Developing a gender policy, addressing gender issues in the workplace and providing gender training as part of staff inductions by specialised personnel are key aspects of the vertical approach. Examining how each and every policy or project (for example, an expansion of the project's area of operation or setting up a new water filtration plant) affects women and men differently, and ensuring that women do not bear the majority of negative impacts, would be the second approach. Making available gender disaggregated data at every step would be an effective measure of gender mainstreaming in the extractive industries. Daly (2005: 437) prescribes that 'a mix of equality approaches'—equal opportunities, positive action and mainstreaming—can and should evolve simultaneously'.

This chapter is the first of 12 chapters written by key scholars and professionals in the field. The second chapter, 'Modernity, Gender and Mining: Experiences from Papua New Guinea' is by anthropologist, Martha Macintyre. Macintyre explores the complicated overlap between the capitalist modernity that is symbolised by the large mining industry and the traditional gender ideologies that are embedded within culture, reflecting on her long engagement with mining projects in Papua New Guinea (PNG). She points out that mining projects—being essentially commercial enterprises meant to raise profits for shareholders—are by nature unlike aid projects, and hence the institutional, economic and social changes unleashed by them are not managed effectively. Macintyre raises the issue in terms of the concept of 'gender mainstreaming', and the ambivalences that are embedded in it. She points to the contradictory stance of large multinational mining projects operating in areas inhabited by indigenous communities characterised by complex gender and social relations. In such circumstances, the companies' notions of corporate social responsibility carry with them the principle of non-intervention in the social and cultural lives of local communities. However, while it is accepted that the mine, in providing employment and developing infrastructure and services, will improve people's quality of life, it remains implicit that cultural practices would neither be impinged upon, nor would the values that sustain customary sociality be eroded. Consequently, the specific disadvantages that women suffer from arise from structural inequities that exclude and disenfranchise women even before a mining project is embarked upon. Indirectly, the chapter highlights the importance of exploring gender-specific disadvantages such as unequal access to and control over resources by women as compared to men. As some of the other chapters also highlight, unequal power relations between men and women are reinforced and accentuated by the large mining projects which establish a 'new social order' in the areas of their operation. Consequently, gender relations in the host communities change in such ways as to further marginalise women from decision-making power.

The third chapter, 'Bordering on Equality: Women Miners in North America' by Laurie Mercier shows how women historically resisted the masculinism of the mining industry and found their way into the male-dominated occupation. Mercier shows how gender ideologies take root and spread across time, space and cultures. The examples from North America of the reactions of male employers and co-workers as women challenged the barriers to mining work are applicable to many other contexts of large and corporatised mining projects. Mercier also discusses how women resisted and persisted in their efforts to break down rigid ideas of gender codes and gained more comfortable lives for their families. Above all, she explores how gender relations change as male miners voice greater support for affirmative action quotas. She also splits open the imagined unitary category of 'women', by showing how much of the resistance to women's employment came from male miners' wives, who were hostile to the idea of employing women in mines. Again, race makes itself visible in her analysis as she notes that many women and men from the black communities supported the hiring of women. Although thanks to the efforts of early feminists women are no longer completely 'hidden' from the history of mining (Burke 2006), Mercier's chapter brings home the importance of understanding history and social context in developing a fuller appreciation of gender in mining.

The fourth chapter, 'Sex Work and Livelihoods: Beyond the "Negative Impacts on Women" in Indonesian Mining' by Petra Mahy explores how one of the main planks of conventional 'victimology' around gender and mining can be rethought. Sex work in mining towns has been a vexed issue where varying perspectives reveal one's moral and/or feminist viewpoint. Women sex workers in the 'negative impacts of mining' literature are portrayed either as victims or heroines, reflecting a futile debate around forced work versus voluntary work.

One of the ways to resolve the debate, she suggests, is to apply some of the more recent theoretical approaches to sex work around large-scale mining operations. This would encourage a rethinking of the roles of women sex workers not generally as victims but as part of the growing and diverse populations around mining towns, as part of the large service industry and as active economic agents. She gives an example from her fieldwork in East Kalimantan in Indonesia, where the operations of a large modern coal mine (KPC) have completely changed the social and cultural fabric in recent years.

The fifth chapter, 'Experiences of Indigenous Women in the Australian Mining Industry' by Joni Parmenter describes the experience of hiring aboriginal women in a large-scale mining project, called Century Mine, located in northern Australia. For many of the women who live in remote communities, this was their first experience in mainstream employment, but it was also an environment heavily dominated by non-indigenous and mostly male workers. The chapter also helps to establish the complexities of gender that lie beyond just sex, and highlights the complex and critical issue of race in gender and mining. Indigenous women and women of colour have argued that 'western' feminists have been a part of the systemic oppression that has disempowered them. Within the dominant 'western' society in Australia, Aboriginal women suffer from a dual oppression from patriarchy and colonialism, and are a double minority. Such women are both indigenous and female, and when hired to work in a very unfamiliar environment such as those in large mechanised mines, they are experience both sexism and racism. Thus it may not be theoretically robust to talk of 'women' as a homogenous and unitary category. A gender perspective would also highlight that Aboriginal women only receive poorly-paid employment in the industry with little scope for career improvement, that is, for their instrumental value as labouring bodies. Parmenter highlights that while the employment of Aboriginal women in the Australian mining industry has contributed to economic gains and empowerment, the challenges faced by indigenous women have been seriously compounded due to additional familial and cultural responsibilities. Employment alone, therefore, is not the panacea towards women's empowerment and gender equality.

A series of subsequent chapters follows, giving examples of what is being done and with what results towards gender equality. The sixth chapter by anthropologist Ciaran O'Faircheallaigh, entitled 'Indigenous Women and Mining Agreement Negotiations in Australia and Canada', exemplifies how indigenous women can be integrated in community consultations from an early stage. Negotiated mining agreements are increasingly becoming more important in Australia and Canada and the benefits such negotiations bring are far from even or gender equitable. These negotiations assume that 'the indigenous community' or 'the indigenous group' is the primary unit of analysis and hence women are marginalised from the very inception of a mine project. With a number of examples, O'Faircheallaigh shows that agreements work better when indigenous women play an active role in the negotiations. He expands the definition of the term 'negotiation' to include activities carried out between the company and the community, and shows that indigenous women play a more complex and substantial role than previously assumed by the literature. Moreover, the specific role played by indigenous women depends on the specific structures used to prepare for, oversee and undertake the negotiation of agreements, and to conduct ongoing processes. Not only has O'Faircheallaigh opened up a new path of research for feminist scholars, his chapter has significant implications for policy makers. For example, while it was the law that created an enabling environment for indigenous groups to negotiate, the participation of women in the cases studied by him were made possible by more gender-specific and conducive cultural factors. It may be possible to enhance these factors so that indigenous women are able to take part in negotiations to ensure that their needs and interests are represented and met by the large mining companies.

This chapter is followed by my own case study of how gender can be integrated within community development projects. Entitled 'Gender-Based Evaluation of Development Projects: The LAST Method', the seventh chapter is based on my experience on a research project (with KPC) in Kalimantan, Indonesia. It shows how the monitoring and evaluation of development projects initiated by mining company 'community development' departments can be performed on a participatory basis. It also shows that such 'asset-based' community evaluations reveal how men and women of different economic groupings are affected differently. While some men more enterprising have tended to benefitted more, with a clearer gender focus, these projects can also be used as key vehicles to empower women. Such gender focus would involve, among other things, listening to the interests of women and supporting and strengthening women's networks and activities.

Following this point, Ana Maria Esteves discusses another area of positive intervention so far neglected in community development projects supported by mining companies. Her perspective has emerged from her experience of working in Ghana, and is presented in the eighth chapter entitled 'Women-Owned SMEs in Supply Chains of the Corporate Resources Sector'. She shows that the integration of local small and medium enterprises (SMEs) in the supply chains of large mining companies can build the capacity of women entrepreneurs and also enhance developmental benefits to local communities. Citing the Ahafo linkages program as an example, Esteves reinforces how support to women entrepreneurs in obtaining procurement and supplies locally can benefit women. Again, this case has implications for community development work by mining companies. Within the resources sector, the measurement of linkages has remained a prominent gap. In their calculations, economists assess the purchase of goods and services in terms of their economic contributions to regions as a whole. However, these activities provide opportunities to plough back money into the hands of women who might spend it more wisely than men. Above all, support to women-owned SMEs in mine-affected regions can also be a starting point to go deeper into the community to address inequalities in power-sharing within the households. Intra-household (mal)distribution of power and resources has been the key area of feminist research on gender and development, and scholars have maintained that women's empowerment would begin in the home. Thus, Esteves' chapter provides for an area of further research.

In the ninth chapter, 'On the Radar? Gendered Considerations in Australia-Based Mining Companies' Sustainability Reporting, 2004–2007' Sara Bice reveals that Australia-based mining companies are devoting more attention and funds to reporting sustainable development. But then Bice raises the question: how do they report gender issues—under the auspices of 'sustainable development' or 'corporate social responsibility' programs? To examine this problematic, Bice uses discourse analysis of the contents of sustainability reports. She argues that the paucity of gender indicators within commonly used voluntary reporting initiatives (such as the Global Reporting Initiative or GRI) contributes to a cycle wherein gender does not make it on to the sustainability program agenda. She notes that it is vital that awareness of gender issues be stretched beyond employment concerns to incorporate gender impacts, which draws attention to the need for further refinement of popular reporting frameworks, such as the GRI.

The next two chapters deal with the other end of the mining and extractive industries spectrum, covering informal, artisanal and small-scale mining. Compared with formal mining, ASM is not only a repository of poor people, it also represents the area where women's labour in mining is concentrated. Women are employed in very large numbers in ASM for a number of reasons that vary from context to context. In general, women are in informal mining because they may be excluded from the better-paid jobs in large-scale mining. Deepening rural poverty, the need to secure cash incomes, male outmigration from rural areas and the realities of women's critical roles in ensuring household food security are factors as important as tradition or generational continuity (Lahiri-Dutt 2008). Trend reports from the mineral-rich tracts of almost all poorer countries reveal that the numbers of women in ASM are increasing. ASM itself is poorly understood and encompasses a diverse range of mining practices. Feminist research and critiques of mining so far have neglected hese mining practices, which comprise a critical element in the livelihood baskets of poor families trying to ensure survival under difficult conditions (see Lahiri-Dutt forthcoming).

Some conventional feminist approaches, such as purely class-based analyses, would be difficult to apply in explaining gender roles and relations in ASM communities. While work is underway to introduce means to accommodate ASM under new mining laws by national governments undergoing economic and mining sector reforms, a lot more work is being undertaken by civil society organisations in some places. International non-government organisations (NGOs), as well as major donors, are working in collaboration with large mining companies which are being sensitised to ASM's importance as a means for survival for the local poor. Pact is one such international NGO operating in areas ridden by ethnic conflict. The tenth chapter 'Towards Post-Conflict Transition: Security of Artisanal Mining Women in the Democratic Republic of Congo' results from Rachel Perks' long-term involvement with Pact in the conflict-torn areas of the country. The chapter shows how critical it is to rebuild and value the roles of women in bringing about a peaceful future and economic revitalisation for the Congo (DRC), where informal mining provides a livelihood for a large number of people. Her chapter shows that investment in women artisanal and informal miners is more likely to increase broader social capital due to their role as sole income providers in many communities, and their relationship with a more traditional and stable rural life. Thus, she concludes, gender issues should be at the forefront of social development initiatives in order to enhance social stability, and development objectives and work—whether in public-private partnerships, NGO programs or UN projects, or concession-mining—must be aligned to a commonly agreed framework to increase their potential for impact.

In Mongolia, the transition from a socialist state to a market-based economy that relies heavily on revenues from resource extraction has fuelled informal mining. Diminishing employment and livelihood opportunities in rural areas, as well as a series of severe winters that resulted in rising commodity prices saw an enormous increase in people involved in ASM within a few years. It was clear that poor people were resorting to ASM as a means of survival. In the eleventh chapter, entitled 'Artisanal and Small-scale Mining, Gender and Sustainable Livelihoods in Mongolia', Bolorma Purevjav outlines the gender issues in ASM in Mongolia and outlines how the Mongolian government, in collaboration with an international donor agency, intervened in ASM communities to improve their livelihoods. Mongolia is an interesting case because the country experienced a phenomenal growth in ASM as the new state embraced liberal principles of economic development. As a large number of nomadic herders began digging up the grasslands, women shouldered a large part of the work required in the ASM processes as well as chores at home. Purevjav builds a case for developmental interventions to improve rural women's livelihoods rather than delegitimising them. Such interventions need to be multifaceted in their approach, addressing social, environmental, legal and economic issues, and she cites ongoing work to reinforce her point.

The twelfth and final chapter, which I co-authored with Gill Burke, entitled 'Gender Mainstreaming in Mining in Asian Countries: Strengthening a Development Perspective', builds and strengthens a development case for gender in mining as against the business case. The argument we have made is that while there has been mineral boom in Asian countries—with generally low scores against the Gender Development Index—this boom may lead to negative impacts on women. The expansion of mining also provides an opportunity to pursue a more gender-sensitive and socially-just development in these countries. Asia has a long history of mining and Asian women have a long history of participating in productive labour, although many new foreign-based mining companies embark with their notions of traditionally submissive, home-based 'Asian women' who have never been involved in 'non-traditional' areas of work. An understanding of the history of women's productive labour and contributions to mining is essential and such knowledge would be helpful for developing gender mainstreaming policies in mining initiatives in Asia.

To conclude, I would point out that there are *five* (often overlapping) major areas of concern in trying to engender the field. To embed gender-sensitive practices and increase levels of gender equity in the mining sector as a whole, the first area of concern is *policy evaporation*. One reason for less attention being paid to gender issues in the mining sector is the assumption that engendered participatory processes will automatically feed into industry policy. While there have been some efforts to ensure community participation and engagement, these processes have yet to become gender inclusive or sensitive to key gender issues. So far, there has not been any clear-cut prioritisation of gender issues on the sustainable development agenda of mining companies. This neglect leads to mining company programs being either gender-blind or gender-insensitive, marginalisation of gender issues in sustainability programs and little or no reporting on gender actions. In various reporting initiatives, gender is not even considered an indicator that is closely related to sustainability. If at all it receives a mention, gender is generally treated as a 'soft issue'. And while the incorporation of 'women' in the GRI's descriptions of community engagement indicators is welcome, it is still not theoretically robust or broad enough to include within its gamut relations between women and men which are dependent on sex, class, race, ethnicity and age. The dilution of a gender focus—due to the lack of clear cut gender indicators—can be described as 'policy evaporation'. Such an evaporation of policy can also occur due to the fear of a backlash; for example, when community relations staff retreat from focusing on gender issues over concerns that all community development work will be reduced to a gender strategy. Evaporation can also be due to the reluctance of mining companies to interfere with 'local culture'—as if a culture is a static and unchangeable object, and as if the company's very presence has not already triggered sufficient cultural change to affect gender roles within the community. To prevent policy evaporation, the international mining community must mainstream gender as a critical aspect of the reporting process.

The second area of concern is the *conceptual confusion* that arises when the outdated WID approach continues to be used, instead of the theoretically more sound GAD approach. WID perspectives are still entrenched in even the most well-intentioned publications in the extractive industries and mining sector. For example, a look at community development projects and strategies would reveal how they continue to target a few 'female' problems in isolation, such as girls' education, women's reproductive health problems or domestic violence.[5] A GAD approach would involve the analysis of inequalities between men and women and then proposing measures that can address those inequalities. Such projects would also involve men so that they too help, support and advocate gender equality. Feminist researchers on mining need to find allies within governments and the industry who can champion gender issues. There is no doubt that women are marginalised from decision-making processes within the mining industry. Further, although it is important to increase women's numbers, simply adding women to planning and decision-making bodies would not address gender issues. As the chapter by O'Faircheallaigh has shown, within mine-impacted communities, women can be involved in crucial areas of participatory negotiations and consultations for better results. In the course of our project experience in Indonesia, we observed that more money remained within the family and was spent on the creation of assets when women were part of consultations involving compensation for land.

The third area of concern is the *mining culture* that includes the attitudes and beliefs of the main actors in the industry. The prevailing 'macho' mining culture hampers the use of a 'gender lens' in areas such as staffing patterns and procurement, as well as community development work. An essential ingredient for effective gender mainstreaming requires staff with the knowledge, skills and commitment to address gender issues in their workplace. So far, the mining industry has seriously lacked strength in these areas. In many companies, no specific gender and equity policies exist yet; the staff induction process does not involve training in gender and equity issues and staff are not systematically exposed to the importance of gender issues. Gender, in general, is subsumed within the broad gamut of 'social' issues. Usually, the work of community relations departments involves a broad range of tasks that may include external relations

---

5   Diane Elson (1991: 1) noted how the WID approach facilitates the view that 'women' as a general category can be added to an existing approach to analysis and policy and that this will be sufficient to change development outcomes so as to improve women's position. It facilitates the view that 'women's issues' can be tackled in isolation from women's relations with men. It may even give rise to the feeling that the problem is women themselves, rather than disadvantages faced by them; and that women unreasonably ask for special treatment rather than the redressal of injustices and the removal of distortions which limit their capacities.

and publicity, in addition to community development. These departments are often 'ghettoised' within the organisational hierarchy and are poorly staffed with relatively low budgets. Even if community development professionals in the industry are committed to gender equity, policy commitments generally fail to be implemented and thus have limited, if any, impact.

The fourth area of concern is the role played by *national laws*. Besides industry culture, laws relating to work and employment in individual countries can play a dual role in hindering gender equity. In countries like the USA and Canada, Equal Employment Opportunity (EEO) and anti-discrimination legislation can create the illusion that gender equity exists in all spheres of life simply because of the existence of such laws, as well as women's advocacy groups and associations. In other (particularly less-developed) countries, laws relating to women's economic citizenship remain discriminatory. For example, the officially recognised 'Head of Household' in some countries such as Indonesia must be a man—the father, brother or husband of a woman—who controls economic decisions, such as those relating to cash compensation for land during mining negotiations. In instances where mining jobs could be offered as compensation for loss of land, discriminatory laws may sometimes exclude women from taking advantage of the economic opportunities created by new mining projects. Described broadly as 'protective' legislation and initiated largely by the International Labour Organisation (ILO) in response to extremely poor working conditions in the early twentieth century, they prevent women from engaging in particular types of hard labour—at night, or in underground mining. In the Indian context, for example, I have shown that the very laws that restrict women's workforce participation also disadvantage them in the job market instead of protecting them, and push women into more marginal and less secure forms of work in informal mining and quarrying (Lahiri-Dutt 2006). Access to all kinds of safe and decent work is the right of both women and men; there is no reason why women in particular would require a special letter of permission from their fathers or husbands allowing them to work in mines (as is the case in Indonesia). Further, the existence of laws tends to give the impression that there are adequate measures for achieving gender equality, potentially resulting in other actors feeling relieved of their obligations and responsibilities to actually uphold women's rights. For example, many countries have constitutions that appear as 'woman-friendly' documents, but in reality, women's rights (such as equal inheritance and land ownership) are not realised in daily life. Once the extractive industries mainstream gender within the sustainable development framework, it will be easier for them to lobby national governments on women's right to mining work.

Finally, as women enter employment in mining, we need a better understanding of the *mining organisation as a workplace*. This includes not only offices but also 'field jobs' in operation areas such as the pits. Workplace issues need to be seriously addressed by the mining industry, which needs to develop a culture of gender equity. While there has been a mushrooming of 'toolkits' and guidelines in recent years, no gender-audit, gender-budgeting, gender-based occupational health and safety rules, or measures to stop sexual harassment and bullying by men, have been proposed. The number and position of women within the organisational structure relative to men are still lower, and the gender wage gap and other such indicators of inequality provide stark evidence of women still concentrated in lower paid jobs within the industry. Again, in dealing with such workplaces, instead of a 'women-only approach', a gender lens would be far more useful. To give just one example, efforts by mining companies to increase the level of employment of indigenous women need to be coupled with measures to ensure these women are given the opportunity to develop their career prospects and rise within the organisation to positions of power and influence. The industry also needs to seriously consider the principles behind diversity; the reasons why it must create an enabling and empowering workplace for women are not clear yet. Some arguments that are widely offered apparently in favour of women's employment can actually be based on biological essentialism and, hence, retrograde. They can lead to further stereotyping of men and women and erosion of gender equity. A classic example of gender stereotyping in mining which ultimately constrains women's career prospects is the use of arguments such as, 'women are safer (in their operational behaviour)', 'women take fewer risks (in driving the trucks)', 'women care for machines (like their children)' or 'women are more docile workers (and do not form unions)'. A critical reflection on such statements would be enough to show that they threaten to mobilise gender for all the wrong reasons in mining—that women should only be hired because they are more compliant or cost the company less. These unverified statements can easily lead us into the trap of biological essentialism that views all women as mothers and carers, and reinforces predetermined notions about who they are and what they should be like.

In mining, feminist research and praxis clearly still has many important hurdles to jump, and there is no doubt that new challenges arising with every positive step. However, none of these challenges is insurmountable. A critical mass of knowledge and expertise now exists, and what is more, this body is growing with contributions from all disciplines and experts from wide-ranging practices, as evident from the contributions in this volume.

I encourage the reader to see this volume as a continuation of consistent efforts by a number of actors to draw attention to the issues surrounding gender and mining. It follows civil society initiatives such as the 'Tunnel Vision' conference organised by Oxfam Community Aid Abroad in 2002, and various international gatherings of the International Women and Mining Network. In 2004, I organised the 'Pit Women and Others' international workshop at the ANU to discuss women miners in developing countries, and jointly edited a book with Martha Macintyre (2006). The World Bank had also successively sponsored 'Women in Mining' conferences in Papua New Guinea in 2003 and 2005. I am delighted to be able to present some of the outcomes of the 'Mining, Gender and Sustainable Livelihoods' workshop to a wider international audience through this publication. While the papers presented here reflect the great diversity of mining practices across Asia and the Pacific region, the contributors share a common purpose—to critically reflect on the lasting gender inequalities within the mining industry in its many areas of operation, and to explore ways in which changes could be initiated. It is not just another volume; it is an outcome of a long process of struggle by a number of agents.

It is a privilege that we can discuss ways and means to positively affect the lives of women and men in mining regions. We can share ideas that can create conditions for women to be empowered, and for men to support such empowerment so that together they can achieve gender equality. Recognising this privilege helps us avoid the traps of tokenism; we all bear a responsibility to ensure, challenge, advocate and insist that all development strategies are gender sensitive. In order to achieve gender equity we cannot give up until it happens. Our privilege also comes with a sense of engagement with and duty to promote and respect the rights of men and, in particular, women, enshrined in international conventions and agreements that have been endorsed by the world and that are generally guaranteed in the constitutions of countries around the world.

References

Blaikie, P., T. Cannon, I. Davis, and B. Wisner, 1994. *At Risk; Natural Hazards, People's Vulnerability and Disasters*. London: Routledge.

Burke, G., 2006. 'Women Miners: Here and There, Then and Now.' In K. Lahiri-Dutt and M. Macintyre (eds), op. cit.

Chambers, R., 2008. *Revolutions in Development Inquiry*. London: Earthscan.

Daly, M., 2005. *Gender Mainstreaming in Theory and Practice*. Oxford: Oxford University Press.

EC (Empowering Communities), 2007. 'Creating Empowered Communities: Gender and Sustainable Livelihoods in a Coal Mining Region in Indonesia'. Viewed 28 July 2010 at http://empoweringcommunities.anu.edu.au/index.php

Elson, D., (ed.), 1991. *Male Bias in the Development Process*. Manchester and New York: Manchester University Press.

Eveline, J. and C. Bacchi, 2005. 'What are we Mainstreaming when we Mainstream Gender?' *International Feminist Journal of Politics* 7(4): 496–512.

Gujit, I. and M.K. Shah, 1999. *The Myth of Community: Gender Issues in Participatory Development*. London: Intermediate Technology Publication.

Lahiri-Dutt, K., 2006. 'Introduction: Where Life is in the Pits (and Elsewhere) and Gendered.' In K. Lahiri-Dutt and M. Macintyre (eds), op. cit.

———, 2008. 'Digging to Survive: Women's Livelihoods in South Asia's Small Mines and Quarries.' *South Asian Survey* 15(2): 217–44.

———, forthcoming. 'Digging Women: Towards a New Agenda for Feminist Critiques of Mining.' *Gender, Place and Culture: A Journal of Feminist Geography*.

Lahiri-Dutt, K. and M. Macintyre (eds), 2006. *Women Miners in Developing Countries: Pit Women and Others*. Aldershot: Ashgate.

Maguire, P., 1996. 'Proposing a More Feminist Participatory Research: Knowing and Being Embraced Openly.' In K. de Koning and M. Martin (eds), *Participatory Research in Health: Issues and Experiences*. London: Zed Books.

McIlwaine, C. and K. Datta, 2003. 'From Feminising to Engendering Development.' *Gender, Place and Culture* 10(4): 369–82.

Razavi, S., 1997. 'Fitting Gender into Development Institutions.' *World Development* 25(7): 1111–25.

———, (ed.), 2009. *The Gendered Impacts of Liberalisation: Towards 'Embedded Liberalism'?* New York: Routledge.

Reeves, H. and S. Baden, 2000. *Gender and Development: Concepts and Definitions*. Brighton: BRIDGE Institute of Development Studies.

UNDP (United Nations Development Programme), 2007. *Gender Mainstreaming in Practice: A Toolkit*. New York: United Nations (3rd edition).

Valdivia, C. and J. Gilles, 2001. 'Gender and Resource Management: Households and Groups, Strategies and Transitions.' *Agriculture and Human Values* 18(1): 5–9.

Walby, S., 2005. 'Introduction: Comparative Gender Mainstreaming in a Global Era.' *International Feminist Journal of Politics* 7(4): 453–70.

White, S., 2006. 'The Gender Lens: A Racial Binder?' *Progress in Development Studies*, 6(1): 55–67.

2. Modernity, Gender and Mining: Experiences from Papua New Guinea

Martha Macintyre

Introduction

While ideologies of human rights, gender equality and the elimination of discrimination underpin most large corporations' employment policies, counter ideologies of gender difference often prevail in practice. Based on research on two major mining projects in Papua New Guinea (PNG), this chapter will explore the ways in which modern western and contemporary Papua New Guinean ideals of femininity, maternal responsibility and the family converge to exclude women from equitable treatment in employment. The chapter constitutes personal reflections and a critique of mainstreaming policies as inadequate to the task of ensuring gender equity in the context of mining in the developing world.

The mining projects where I have undertaken research in PNG have been operated by large international companies whose websites and stated policies proclaim commitment to principles of human development; to social and economic sustainability when working in developing countries; equality of opportunity in the workforce; equal pay for equal work; localisation of the workforce; and provision of training and support for the pursuit of further education or qualifications for local people. In each place, departments devoted to implementing social responsibility match this public commitment. Human resources sections, community affairs (now almost invariably featuring 'social sustainability' in their titles or mission statements) and specially-designated positions within these departments attend to projects that are aimed at gender equity, women's activities in the community and sometimes youth affairs and recreation.

In PNG, mining is one of the major industries. It is seen by many local people as the most likely means of improving standards of living, providing employment and economic changes that people associate with development and modernity. While NGOs and international providers of government-to-government aid inevitably stress gender equality as an aim in any project within PNG, the most significant economic developments come through industrial projects (almost all resource-extractive) and within these projects, gender inequality is not salient. In fact, one could argue that it is well and truly 'mainstreamed'. Gender inequality is meant to be achieved by policies that are non-discriminatory rather than through policies and programs that redress the imbalances and enable women to participate equitably.

If we take the definition of gender mainstreaming from the United Nations literature (UN ECOSOC 1997; UNDP 2005) it requires a full assessment of the implications for men and women of any project so that the interests of men and women are integrated into the design and implementation of aid delivery, economic development programs and projects aimed at improving the political and social participation of people. This is meant to ensure that '… men and women benefit equally and inequality is not perpetuated. The ultimate goal is to achieve gender equality.'

Gender mainstreaming is readily comprehensible in the context of aid projects where the express aim is a managed intervention aimed at effecting change—institutionally, economically or socially. But when the project is a commercial enterprise, such as a mine, the transformative project is less clear. The notion of corporate social responsibility, for instance, often carries with it ideals of 'social sustainability' and non-intervention in aspects of the social and cultural lives of local people affected by economic change. The requirement of social impact analysis in some countries rests on an idea that the mine, in providing employment and developing infrastructure and services, will improve people's quality of life but should not impinge on cultural practices in ways that erode or destroy values that sustain customary sociality. Often this is interpreted as maintaining the status quo, as if 'customary ways of life' can be insulated from the upheavals generated by economic transformation.

To give examples from two mines: the influx of money from compensation, royalty payments and wages at Porgera in Enga Province enabled those men who became wealthy to pay high bride prices and marry multiple wives on an unprecedented scale. The social effects of this new practice have had destabilising effects on marriage, household composition and social harmony (Bonnell 1999; Golub 2006). Many women perceive this as having contributed to a decline in their status generally and to the erosion of customary ways of negotiating marriage. The abandonment of older wives and increasing number of wives taken from other tribal groups is seen as a factor in increased incidence of domestic violence and tensions with neighbouring groups (who view marrying-in as a strategy for gaining access to benefits). Because men have greater access to financial benefits, especially employment, women sell sexual services—with the consequence of a very high incidence of sexually transmitted disease, including HIV/AIDS (Hammar 2008). These social effects were predictable and are attributable to economic and social changes brought by the influx of cash and the extension of its use in sexual relationships. However, it could be argued that measures aimed at increasing financial benefits to women, or having a rigid policy of gender equity in employment would have been interpreted as an assault on customary gender relations. It is also highly unlikely that people, both male and female, would support policies that require equal numbers of male and female employees.

In Lihir, when the groups of landowners set up bank accounts for the receipt of compensation and royalty payments, no women were given authority to control the accounts. An attempt by the company to insist that women landowners be named on the lists and have some control over moneys was rejected by both men and women, who viewed men's privilege in this domain as an extension of customary control over a lineage's wealth. Sensitive to accusations of cultural insensitivity, reluctant to interfere in ways that could be viewed as interfering in private financial matters within families and having no legal authority to determine how each local group deals with its money, the mining company could not respond to remedy gender inequities.

There are problems with the level of generality in notions of mainstreaming gender in mining. What does it mean to women that they benefit 'equally' from a mining project? That they have equal access to any jobs? That there are equal numbers of men and women employed? That the distribution of wages is such that men and women have equal share? The infrastructural and service benefits such as roads, hospital, schools and telecommunications can, and often, benefit women at least as much as men. Reticulated water and electricity can reduce the amount of time and effort that women expend each day transporting water and fuel. Such services can also improve health, so that respiratory illnesses associated with cooking over open fires are reduced, or a clean water supply improves hygiene and reduces waterborne diseases. Being able to wash clothes in a laundry rather than on a mosquito-infested riverbank might also mean that women are less exposed to diseases such as malaria. A gender analysis of proposed changes would identify the gender dimensions of any proposed change in order to ensure that communal benefits were spread equitably. As women's needs and aspirations in a community are often quite different from those of men, undertaking a gender analysis that identifies these distinct needs and establishing projects that ensure they are met with would seem adequate to the situation. It could also ensure that women's needs were met, an end not that is not secure by applying gender mainstreaming.

In my experience miners view such benefits ambivalently. They are a cost at a time when the profits have not begun to roll in. Mining companies recognise that electricity, water supplies, roads and hospitals are necessary for their own enterprise and as such can be extended to the local community more cheaply than other infrastructure. They are also concerned about support from local people and thus present these as being benefits that are responding to their demands. To some extent this is true—people do want roads and other services.

But they also want housing, school buildings, money in their pockets and a chance to set up business enterprises. Here companies are less generous. They assist in establishing businesses that support their own needs—contractors who can supply goods and services to the mine. They build houses for relocatees, but not for others who simply live in the area. In short, they try to make their contributions to local communities that further their own business interests and help the mine to run smoothly. They meet the demands of their hosts as much as possible to avoid shutdowns and bad press, but these demands come mainly from men. Women rarely present themselves at meetings of stakeholders. If asked what businesses they might be interested in developing, they usually lack any experience that would suggest that they could operate a business that supplies goods or services to the mine. Their absence from negotiations at crucial initial stages means that their involvement in economic development is limited from the outset.

Given the cosmopolitan character of mining companies, they are familiar with the principles of gender equity in the workforce and it is noticeable that in PNG these are generally adhered to as standard practice. Job advertisements are gender neutral. Within the mining sector there are female graduates who work in the professions and technical areas and are paid at the same rates as men in those positions. I have encountered women who are geologists, engineers, chemists, heavy machinery operators, accountants, biologists, librarian/archivists, environmental scientists, computer scientists and human resource managers. During the first few years of working in Lihir, when I was given access to data on applicants for positions, it was clear that with the exception of administrative positions where there was a higher percentage of female applicants, more than 90 per cent of applicants were male. Tracking over the years 1995–1997, it emerged that all women who applied with appropriate qualifications were almost certainly considered or interviewed (Macintyre 2005: 22). During the first five years of the mining project, the highest paid Papua New Guinean employees were predominantly female, with most of them in administrative positions—but there were very few of them and all were subordinate to male managers. During that time there was only one PNG man in a managerial position, all senior managers were expatriate men.

As a transnational industry, mining employs senior people (mostly men) who are able to move from project to project and often from one country to another. The fly-in/fly-out rostering policies for expatriate employees adopted in PNG favour men who are either single or married with a wife and family in their home country. The gender norms that prevail in Australia (where most expatriate managers originate or live) mean that such mining jobs are often ones that women regard as too disruptive of family life. Living on site enables people to have more normal domestic arrangements, but even then the long shifts and cyclical rosters do not accommodate the needs of children (see Pattenden 1998; Yrke 2005). For expatriate fly-in/fly-out workers, the bonus is in the remuneration; as Yrke found in her study, wives accepted the inconvenience and added burdens of being the sole parent most of the time because the husband earned a much higher wage than if he were in an equivalent position close to home. It is in this context that the mining industry's advertisements can appear to be non-discriminatory—they are adapted to the gender ideologies of the industrialised nations in which men are considered primary breadwinners, women take major responsibility for children and home (even when they are employed).

If the employment policies of the mining company reflect their commitment to gender-neutral procedures, those of the local contractors for services displayed prevailing ideas about female roles in PNG. The only company that employed a large number of women was the catering firm contracted to do the cleaning and meal provision. Women cleaned, washed clothes, served food and worked at the reception desks for the camps with low wages.

Early in the construction phase of the mine one of the local contracting companies decided to give women the opportunity to train as drivers of the large trucks and machines used in excavating the mine area and building the roads. They hired a female trainer and encouraged all Lihirian women to apply. Eight women completed training, most of them from Bougainville, and five gained employment. After a year only two Bougainvillean women remained—all the others decided that they would stay at home and avoid the hostility of male employees, husbands and villagers. The resilience of the Bougainvillean women testifies to their long familiarity with mining and their experience of exclusion from the opportunities to work in male-dominated areas. They recognised that wage disparities based on gendered occupations excluded women from jobs where the salaries were commensurate with those of men.

As the conventions of the division of labour that prevail in commerce and industry in PNG have their origins in the Australian colonial system, most local people accepted that women's jobs would be predominantly in the provision of clerical or domestic services. Similarly for the mining company—women worked in clerical positions while men would be concentrated in the skilled and semi-skilled jobs in the pit and plant site. The training reports for the early years of the mining project on Lihir reflect the acceptance of this gendered division by both men and women. Female trainees for clerical work attended courses conducted by women where they learned basic keyboard skills, filing and other office work. They were also given lessons in dress and grooming— learning how to paint their nails and apply make-up. While most did not practice these newfound feminine arts, they expressed delight and enthusiasm at having acquired them. The reason that they did not wear make-up or paint their nails was that people in the village would see these as indicators of their sexual availability. This touches one of the major problems facing women who enter the workforce, namely, that women who move beyond the bounds of home, family or garden are often considered to be actively seeking sexual adventure. This discouraged married women from working in places where men also worked.

In countries such as PNG, where the normal division of labour tends to involve gendered tasks or familial cooperation, adjustment to working conditions that require unrelated men and women to work together can be problematic. Working women described how husbands were suspicious and jealous to the point where they were even violent towards women they suspected were 'too friendly' with their male colleagues. In one discussion about the difficulties women faced, one woman explained that shortly after she was married her husband observed her plucking her eyebrows. He instructed her that this suggested that she wanted to attract other men and was not to continue this practice.

I have been conducting research into the experiences of women in the workplace in PNG for a period of eight years with about 100 surveys completed and 30 in-depth interviews. In particular I have surveyed the work experiences, hopes and aspirations of women who have been educated beyond secondary school. While most of the participants have been working in government jobs—as nurses, teachers and police officers—a significant group comprises women who work in the minerals industry. This group has the highest number of graduates and these women are younger than their public sector counterparts. We might therefore consider them to constitute the next generation of women who are potential leaders. They are characterised by a greater awareness of their capacities as professionals and unlike others they tended to express ambitions that extended beyond being successful in their current jobs. They are slightly more likely on average to control their own incomes and to live independently of other family members.

Yet the difficulties they experience at work are remarkably similar. The two problems reported by almost all women were the difficulties in gaining promotion when men would be working under their authority and the domestic problems with male partners who were jealous and suspicious of their fidelity when they had any men working with them. Seventy per cent of the women mentioned domestic violence as a problem faced by working women. As 20 per cent were single, divorced or widowed, this is an alarming proportion of women who are intimidated by men simply because they work in paid employment. It needs to be kept in mind when we consider the question of the 'empowerment of women'. It also means that simple measures such as having policies of equal pay for equal work or gender-neutral advertising of positions are not strategies that immediately facilitate or encourage equality. They operate within a social framework where there are often rigid gender roles, where women are unaccustomed to working with men, where men expect women to be subservient and women who 'break' gender norms are stigmatised as (at best) sexually available and (at worst) sexually promiscuous.

These problems are not insurmountable, but would require groundbreaking policies aimed at changing the preconceptions of both employers and potential employees. In speaking with women who were employed in lower paid clerical positions about the possibilities of training for positions that they considered 'masculine', many demurred on the grounds that they preferred to work with other women; they were sure that men would object and because they liked the 'clean' environment of an office.

Ideally, gender-mainstreaming policies would require a thorough analysis of the ways that the workforce operates to limit female participation, in combination with a critique of deeply entrenched gender ideologies held by employers, employees and the local community. In order to produce any effective change, there has to be political will and support from all involved. In PNG there is no indication that men in government or at the grass-roots level are prepared to relinquish their power over women's labour and fertility simply to comply with the requirements of a UN convention or the ideals of liberals in much wealthier, industrialised countries.

Similarly, mining companies have been slow to change policies and practices that diminish their profits in western countries—higher environmental standards; improved working conditions and equal opportunity in employment have been fought for in liberal democracies and adherence to higher standards generally has been achieved through legislation. In PNG, where several companies continue with practices that would not be acceptable elsewhere (such as dumping tailings and waste rock directly into rivers), there is no pressure from the government to enforce international 'best practice'.

Mining projects tend to be located in areas where people are subsistence farmers. The aspirations of modernity that people hold during the initial stages of a project rarely include female training and employment as a form of social advancement. Rather, both men and women desire improvements in infrastructure and services—especially health and education—and houses that they associate with urban locations. In both Lihir and Misima, places where I undertook social impact studies before mining began, most adult women wanted a 'western-style' house and expected that the men in their family would be employed, while they would continue to produce garden crops and perhaps have access to cash by selling some of its produce. A few hoped that their educated daughters might find employment, but none suggested that this should be a route to gender equality. These modest hopes have been only partially fulfilled in all areas where mining projects have been established.

A gender analysis of the situation in mine-affected areas would reveal consistent, structural female disadvantage. The forces that sustain this situation are not all exogenous; in fact the most difficult to overcome are probably those deriving from the local customs, and economic and political conditions, that preceded mining. When these combine with the masculine ideologies of mining, whose main interest is profitability, the notion that 'gender mainstreaming' will be acceptable to all parties is simply fanciful.

For example, in PNG there is the problem of the 'double shift' for local women who work. In most countries, even those where women have been in the labour force for generations, the burden of domestic labour continues to be disproportionate. Mining projects occur in rural areas where there is rarely any precedent for employment for local women. In PNG, where gardening and the provision of the family's staple food is part of their domestic duties, many women who work continue to grow gardens for their food supply. In interviews, the overwhelming majority of female employees in Lihir reported that at least one of their days off each week was spent working in their gardens.

Women are expected to continue all the work of being a housewife and mother. None of the mining companies in PNG offers paid maternity leave. In Lihir basic provision was made for women to continue breastfeeding after returning to work. There was a building provided in the mining camp where babies could be cared for by a member of the family (usually a young female relative paid by the mother). It was bare except for a couple of beds and a few chairs. It had running water and sinks. It was surrounded by a wire fence and had no facilities for children to play, It would have failed Australian standards of childcare provision in terms of hygiene, space and amenities. Given the sociable and varied life available to a village infant, this could scarcely be seen as a major contribution to the welfare of mothers or children. There is no workplace childcare at any mine.

Mining companies are male dominated. The boards, management and workforce are overwhelmingly male. The prevalence of fly-in/fly-out arrangements, while minimising some of the adverse social impacts on local populations, assumes that the employee is single or is a man who has a wife who takes care of the home and family. Camp life and facilities cater to men with few concessions to women's ideas of recreation. The bar and snooker tables occupied the whole of the indoor recreation room space in both the camps where I worked.

But in considering the problems faced by women who work in mining I want to stress those home-grown gender ideologies that are reinforced within this masculine workplace rather than being challenged by notions of gender equity and non-discriminatory policies. The reasons are simple and common to many other areas of life. They have been extensively canvassed in documents relating to human rights and development policy and are enunciated clearly in documents such as 'Taking Action: Achieving Gender Equality and Empowering Women', by the Millennium Development Project's taskforce on gender equality (UNDP 2005). The issues identified by a gender analysis are all ones that primarily or exclusively affect women adversely.

In a chapter examining the success and failure of gender-mainstreaming in the context of achieving the Millennium Development Goals, Rekha Mehra and Geeta Rao Gupta assessed numerous NGO and government projects where this approach had been implemented. They noted that one of the factors that limited the success of the strategy was the idea that 'all staff should be responsible for its success' which in practice means that '… when mainstreaming is everyone's task, it can become nobody's responsibility' (Mehra and Rao Gupta 2006: 5). They cite the case of the Dutch government where 'an immediate consequence of the policy's adoption was closure of all gender equality offices at the local level—nobody assumed specific responsibility, procedures did not change and, as a result, gender equality goals can be swept away by the mainstream, instead of changing it.' (ibid.).The obverse can hold true also—when gender is the specific responsibility of a designated person or group within an organisation, then others do not have to ensure that policies are successfully implemented and the 'gender specialists' can be dubbed ineffectual.

But the more telling message that emerges from this document, as from others dealing with gender mainstreaming, is that all the inequalities that are identified are ones where women are the disadvantaged group. This is my main objection to the change in nomenclature. Gender mainstreaming is effectively a way of pursuing goals of gender equality that almost everywhere requires the development of policies and strategies that will improve the status of women. They will also eliminate discriminatory practices that exclude women from equal participation and allow women to benefit in economic and political activities. As this is so clearly the case, why do we not dispense with the verbiage and begin from the outset to deal with these inequities as problems that require solutions that concentrate on women?

A couple of years ago I read Stephen Lewis' (2005) book about women and HIV/AIDS in Africa. I was preparing a lecture on Women and Development in Melanesia and it was about 10 years after I had been one of the four contributing authors to the World Bank Report on Women in PNG (Brouwer et al. 1988). Looking up the UN Human Development indicators I realised how little had changed for women there. Women's lifespans have barely increased; the high rates of female illiteracy, maternal deaths, malaria, tuberculosis and HIV/AIDS tell us that women's lives have not improved much at all. The incidence of domestic violence, lack of women in political leadership and low rates of women in higher education and employment show how far they are from achieving gender equity. In PNG, as in many other developing countries, disparities are so extreme that any attempt to redress the imbalance could only succeed if radical changes in access to education, political participation and employment were enforced through legislation.

The following is a quick gender analysis. The difficulties that women face in getting the education and training that enables them to gain employment in mining projects are multiple and reinforce each other. The reasons for this are complex: a mixture of gender ideologies that are compounded by poverty; a lack of government commitment to the education sector; a subsistence economy that depends on female labour; and a lack of sex education and services for family planning (many secondary school students leave education because they become pregnant). The problems they face when they get there are often even greater: open antagonism; sexual harassment and discrimination in terms of opportunities for promotion, remuneration and in-service training. At home, married women regularly face the jealousy of husbands if they have to work shifts with men and husbands will try to prevent them from engaging in any training that requires them to travel. In short, a gender analysis generates an overview of a situation in which women's disadvantages are constantly compounded by the interaction of factors in the workplace and at home. There is a need for specialised support programs for women and various forms of affirmative action that enable women to participate in the workforce on something resembling equal terms.

'Gender mainstreaming' is an approach that assumes that the conditions for an increase in female participation exist in the female population and can be redressed by adjustments and policy changes that allow women to do such things as enter the workforce. My research and experience of working in PNG over the last 30 years suggest that improving gender equity is a lot more complicated. Whatever gender mainstreaming might be in academic terms, by the time it gets into aid projects or workplace policies it has become 'add women and stir' with nobody prepared to actually do this. If gender equality is to be achieved in a period of a decade, specialised programs for women are needed.

Stephen Lewis' work on HIV/AIDS in Africa provides a discussion of this, which for all its polemic struck me as directly applicable to women's employment in mining in PNG:

> ... gender mainstreaming is a pox for women. The worst thing you can do for women is fold them into the mandates of ... [broader development programs]. Once you've mainstreamed gender it's everybody's business and nobody's business. Everyone's accountable and no one's accountable. I don't know who thought up this gender mainstreaming guff, but I often wonder what their motives were.... Gender mainstreaming might work

if we had what sports and financial enthusiasts call 'a level playing field', that is to say, if there were real equity and equality between women and men. Then gender mainstreaming becomes a way of maintaining that equality. But when you start from such gross inequality, mainstreaming simply entrenches the disparities (Lewis 2005: 125–26).

My experience in PNG has convinced me that if women are ever to operate on a 'level playing field' then specific attention will need to be given to overcoming the structural disadvantages and inequities that exclude and disenfranchise women before any project (including mining projects) is embarked upon. Mining companies are private enterprises and although they might be agents of change, they should not be usurping the role of the government. While they might, for instance, develop training programs aimed at increasing female participation in employment, it is not appropriate that they should be the implementing agents of UN conventions such as that for the Elimination of Discrimination against Women (CEDAW). Programs and projects that are aimed at improving women's education and training, policies ensuring that women have access to services that deal with their reproductive health, legislation that requires the provision of work-based childcare or equal pay—these are matters for governments to establish and enforce. In the current environment, where women's disadvantages are so entrenched, there is an urgent need for policies that are directed exclusively to improving women's lives and opportunities so that they are able to participate in social and economic life on equal terms with men.

Government actions in the form of legislation and enforcement of anti-discriminatory policies can 'level the playing field'. Commercial enterprises can only work in the context of national policies and laws. In the mining industry in PNG the steps that can be taken to improve opportunities and conditions for women will still have to take the form of affirmative action if equality and equity are to be gained within the next decade. At present there are a few aid projects that are working exclusively with women in order to change their status, increase participation in political affairs and engage in campaigns for recognition of their rights, but there is little will within the government to effect such changes. Mainstreaming within an industry will simply mean that gender-specific disadvantage is ignored and that companies continue to make the excuses that 'women did not apply', or they 'cannot work shifts because of domestic responsibilities' or 'there were no suitable female applicants'. In my experience, mainstreaming means that women's specific interests are subjugated to the needs and objectives of a project. In aid projects, the mainstreaming agenda is linked to cost-cutting; restructuring departments so that women's issues are diminished in importance and funding is based on large programs that subsume women as just another factor in the overall project. In the mining industry in

PNG, where there has never been a tradition of even noticing women's interests as an issue until they are manifest as social problems, it means they would be marginalised from the outset and 'disparities would be entrenched'.

References

Bonnell, S., 1999. 'Social Change in the Porgera Valley.' In C. Filer (ed.), *Dilemmas of Development: The Social and Economic Impact of the Porgera Gold Mine, 1989-1994.* Canberra: Asia Pacific Press (Pacific Policy Paper 34).

Brouwer, E. B.M. Harris, and S. Tanaka (eds), 1998. *Gender, Economic Development and the Status of Women in Papua New Guinea.* Washington DC: World Bank, East Asia and Pacific Region.

ECOSOC (Economic and Social Council of the United Nations), 1997. 'Mainstreaming the Gender Perspective into all Policies and Programs in the United Nations System.' Geneva: United Nations.

Golub, A., 2006. 'Who is the "Original Affluent Society"? Ipili "Predatory Expansion" and the Porgera Gold Mine, Papua New Guinea.' *The Contemporary Pacific* 18(2): 265-92.

Hammar, L., 2008. 'Mobility, Violence and the Gendering of HIV in Papua New Guinea.' *The Australian Journal of Anthropology* 19(2): 125–250.

Lewis, S., 2005. *Race against Time: Searching for Hope in AIDS-Ravaged Africa.* Melbourne: Text Publishing.

Macintyre, M., 2005. 'Social Impact Report for Lihir, 2004'. Melbourne: University of Melbourne.

Mehra, R. and G. Rao Gupta, 2006. *Gender Mainstreaming: Making it Happen.* Washington: International Centre for Research on Women.

Pattenden, C., 1998. 'Women in Mining: A Report to the Women in Mining Taskforce.' Carlton: Australasian Institute of Mining and Metallurgy.

UNDP (United Nations Development Programme), 2005. *Taking Action: Achieving Gender Equality and Empowering Women.* London: Earthscan.

Yrke, M., 2005. "What You Do is Who You Are": Gender Identity in the Resources Industry. Perth: University of Western Australia (Ph.D. thesis).

3. Bordering on Equality: Women Miners in North America

Laurie Mercier

Introduction

Mining has often been upheld as the most 'masculine' occupation, and as one that traditionally prevents women from entering it, be it in sixteenth century Peru or twenty-first century South Africa. As many scholars have demonstrated, these restrictions are historical and social in nature, constructed to limit women's employment and encourage their reproductive labour at particular points in time. Yet these have been enforced as if 'natural'. In Canada and the US, two decades—the 1940s and 1970s—offer exceptions in women's employment in mining and, therefore, some insights into how women and their advocates might dislodge gender barriers. The Second World War presented a labour shortage that forced governments, companies and unions to recruit women to fill critical mining positions. Later, North American feminist movements compelled governments to open up former male bastions such as mining through Equal Employment Opportunity (EEO) legislation.

This chapter explores the reactions of male employers and co-workers as women challenged barriers to mining work and how they found, kept or were terminated from employment. Structural and cultural resistance appeared unyielding. The questions raised by the chapter include: How did women overcome obstacles to their employment? How and at what times did companies and unions attempt to break down or sustain rigid ideas of gender roles? Why did the number of women miners not increase more substantially over the next several decades?

The title of the chapter signifies at least two kinds of borders. The physical border between nations (in this case the US and Canada) has not historically impeded the movement of capital and ideas about gender. Women in mining have also been positioned on the borders (or barriers) of a male-dominated occupation. Despite legislative efforts such as EEO, most women miners have been discouraged by entrenched cultural resistance to their employment, alongside economic factors including mechanisation, plant shutdowns and declining jobs. These examples from Canadian and US mining workplaces illuminate the difficulties in instituting gender equality, and this historical perspective may suggest some lessons or directions to how to overcome those inequalities in mining worldwide.

Masculinity and Mining: Background

Two important recent, comparative anthologies reveal that gender constructions of work, family and militancy have been central to the masculine world of mining; and that women have played a more critical role in mining enterprises and communities than previously understood. In *Women Miners in Developing Countries: Pit Women and Others* (2006), Kuntala Lahiri-Dutt and Martha Macintyre present a number of essays that focus on artisanal and small-scale mining (ASM). ASM traditionally employed larger numbers of women than large-scale mining; and these women combined their earnings from mining with agricultural subsistence activities. Even when women were excluded from digging underground, they played pivotal economic roles on the surface in transporting and processing the diggings.

With tremendous variety in the world's mining enterprises, how did the occupation become so singularly male-centric? In Gier and Mercier (2006), it is found that gender exclusions emerged at particular historical moments and were hardly uniform across the world. For example, women in precolonial Africa mined below and above ground, while in the precolonial Andes it was considered 'bad luck' for women to work underground although they were central to mining work. Global historical forces of colonialism or capitalism often altered customary gender roles in mining, limiting women's traditional rights over mineral wealth. Attitudes towards women and mining were uneven in industrial societies as well. For example, the British parliament officially prohibited women from underground work as early as in 1842, although women miners continued to work in some regional pits in the early twentieth century. Elsewhere in the British Empire, Indian women laboured in British-owned mines for a fraction of the wages paid to their male counterparts. And even in societies such as Japan, where gender roles were rigidly enforced, surprisingly, women continued to mine underground until the mid-twentieth century.

Despite the variety of places in the world and mining activities represented in these two mentioned works—from gold to coal to copper and tin—a clear pattern emerges that links women's lives across time and space: as the processes of globalisation have expanded into postcolonial spaces, mining has become more masculine. These works conclude that the position of women in mining declines as the operation becomes more capitalised, centralised and mechanised. Whether in Asia, Europe, the Americas or Africa, societies and employers normalised women's exclusion from working underground through an elaborate set of superstitions, beliefs, traditions, sexual metaphors and seemingly 'rational' justifications, which were enforced through legislation or cultural taboos (see Francaviglia 1998).

Since we know that ideas about gender roles are historically and socially constructed, we need to know when, why and how mining became particularly resistant to women working underground. Yao (2006) provides a fascinating glimpse of how the numbers of Chinese women in mining rose and fell with changing political and economic tides. Communist emphasis on equal rights ended Confucian gender restrictions, but after economic reforms, women in China have seen their share of mining employment decline since the 1990s. In the 1950s and 1960s women miners were glorified as 'Iron Girls', but in reality they still represented a minority of workers underground. Some high profile accidents in the 1980s led to the 1992 regulation called the 'Provision for the Protection of Working Women', which prohibited the employment of women underground. With a new emphasis on markets, women workers were seen as more expensive with their maternity leave provisions, and less reliable than men. As a result of these restrictions, Chinese women have moved figuratively underground, working illegally in mining activities and losing prior benefits.

This is but one example that clearly illustrates the vacillation of employers and the state in either embracing or rejecting women miners, almost always tied to structural conditions. Companies or nations saw their own interests tied with patriarchal male miners and reinforced gendered practices, although they alternately viewed women as assets or liabilities in an effort to control labour. Companies often encouraged marriage and constructed family housing in order to sustain a more docile workforce; at other times, they tried to limit the number of women in a mining camp. Although men came to mostly dominate mining, women's reproductive and domestic work were essential to the industry. As a US Women's Bureau study concluded in the 1920s, miners' wives were of 'peculiar industrial and economic importance' to keep miners in the region (Lahiri-Dutt and Macintyre 2006: 6). Moreover, women's waged and unwaged work played a critical role in the basic economy of the industry. Women created economic niches through direct relationships to mining in surface operations that admitted them—along with entry in brothels, taverns and cafes, households and other businesses that maintained male miners.

Since mining has historically excluded women from employment, much research has focused on ways that women, through their reproductive labour, supported the industry; and, in their militancy, helped male miners through strikes and other labour actions. Despite company hopes that they would 'settle' and make a more dependent male workforce, women often became the more militant community members in demanding better wages and living conditions.

In fact women often manipulated prevailing gender assumptions to more effectively and physically assert the goals of labour strikes. Women took over when men were restricted from participating in public demonstrations through injunctions, or experienced military and police violence. Labour movements preserved the ideals of female domesticity and the male worker as the head of household, but women creatively exploited these assigned roles to pursue their own interests and their own forms of protests. Thomas Klubock (1998) found in Chile's El Teniente, for example, women used the 'family wage' ideal to claim rights to their husbands' wages and benefits as well as work with them to extract economic and social concessions from Kennecott Corporation.

Abundant examples from the world's mining communities reveal how women exaggerated gender claims in solidarity for what they viewed as family and community, not just union, efforts. As Janet Finn (1998) documents in Chuquicamata, Chile, when the company tried to crush a strike, groups of women hid lunch buckets and even seized and dressed strikebreakers in women's clothing to ridicule their lack of manhood in failing to support the strike.

In the process of supporting men's labour rights, women often came to contest the gendered rules for protest and question their own roles in unions, families and communities. Because these moments of protest often appeared as much about challenging patriarchy as capitalism, male miners and unions were not always supportive of women's independent militancy when they performed more traditional support roles during strikes. Despite women's proven critical role in labour action, male miners often 'ordered wives back to the kitchen', as June Nash (1979) discovered in her study of Bolivian tin communities. But women and men repeatedly struggled over and renegotiated those gender roles.

Some of the greatest but less visible struggles occurred when women sought access to the mines that would allow them the opportunity to earn livelihoods equal to those of men. A historical approach allows us to see how gender conceptions are fluid and intersect with other identities such as class, race, culture and nation, as we examine how a minority of women boldly asserted their economic and social rights to do 'men's' work.

Women and Mining: North American Context

Two types of women miners have drawn historical attention: the intrepid individual prospectors of different class backgrounds who prospected the Rocky Mountains West from New Mexico to the Yukon in the nineteenth and early twentieth centuries; and the women who laboured underground for family survival, almost always in small enterprises, primarily coal, rather than hard rock, and alongside husbands or fathers.[1]

Although these examples are noteworthy in the sense that they challenge the assumptions about women's underground mining work, they contribute to an exceptionalism that may hinder a closer look at the gendered structural exclusions in corporate mining, which has dominated the industry in North America since the late nineteenth century. Since we know that women have been employed in smaller mine operations, and legislation and custom even allowed them underground, particularly when working with family members, it seems to me that the chief question to pursue, then, is why women could not work in larger mining operations?

Mining has been part of a larger culture of masculinity that infused industrial work. From the late nineteenth through to the late twentieth centuries resource-based industries—logging, mining, agriculture, fishing—distinguished the gendered and racialised character of work in the US and Canada. Even though women and coloured people worked in these industries or the service sectors that supported them, narratives of regional work reinforce the concept of a white, male 'wageworkers' frontier' (see Schwantes 1979, 1982, 1988). These occupations elicit images of tough, masculine outdoor work and independence. They have also resisted hiring women and certain ethnic groups except when labour demands overwhelmed the exclusionary rigid boundaries they erected. The white male breadwinner ideal and the reputed toughness of the work that supposedly discouraged women from employment often disintegrated when labour markets expanded, when families required multiple breadwinners, or when these 'rugged' jobs became seasonal and low-paid, which then made women and people of colour 'ideally suited' for the work.

Industries, unions and male workers perpetuated this social order that reinforced the belief that much industrial work required rugged masculinity. Unions and workers might preserve jobs for white men, but employers could resist improving conditions or providing adequate workers' compensation by exploiting stereotypes that rendered workers tough enough to handle the most dangerous conditions or too rootless and family-less to warrant company-provided protections (Robbins 1986; Forestell 2006).

---

1 For example, Irish immigrant Nellie Cashman, through her lifetime, searched the Klondike of the Canadian Yukon and north of the Arctic Circle in Alaska for precious metals and gained fame as a philanthropist from her mining earnings (Chaput 1996). For the history of women mining entrepreneurs see Zanjani (1997). On women coal miners during the Great Depression and other periods, see Gúerin-Gonzalez (2006) and Moore (1996).

Despite union gains in North America after the 1930s, many mining union leaders and rank-and-file workers continued to believe that the association of masculinity with militancy aided their cause. The International Union of Mine, Mill, and Smelter Workers (IUMMSW), which represented the US and Canadian non-ferrous miners and refiners; and the United Mine Workers (UMW), representing coal miners, championed masculinity in their iconography and rhetorical traditions to assert their fierce independence. The mining industry's contraction during the 1920s and 1930s also reinforced it as a male domain, but the sudden demand for labour beginning in 1940 upset gendered notions of work.

Labour Expansion in the 1940s

Even during the Second World War—when much of Canada and the US opened industrial jobs to women—company, government and union officials sought to preserve men's claims to traditional mining occupations. Montana's war manpower director, for example, claimed that the state needed 'men for the hard, heavy and unpleasant jobs' in mines, mills and woods 'where women cannot be used'. Anaconda Copper Mining Company and Mine Mill union officials agreed that mines, mills and smelters could not employ women because the work required strength and stamina. But physical prowess evidently was not the chief requisite because the company began recruiting retired and disabled men (MHSA 1943). Nancy Forestell (2006: 86–7) charts how gold mines in Ontario resisted hiring women even during the war labour shortages. In other mining centres companies supported hiring women with the understanding that they could free-up large numbers of men on the surface to work underground. But gold mining executives decided not to pursue female workers since the mines employed disabled male workers on the surface, maintaining the gender exclusivity of the industry.

The war years provide ample examples to illustrate the shifting ground of gendered and racialised work categories. In Anaconda, Montana, the Anaconda Company manipulated perceptions of difference to convince union representatives to allow women into its smelter. Appealing to racial prejudices, the management threatened to import African-Americans and Mexicans to fill the labour void, emphasising that they would prefer Anaconda women 'rather than Mexican boys', but the federal government could send 'coloured men' any time. Management and labour then agreed that they would preserve community values that championed white male and female breadwinners. The new employees had to be Anaconda residents, wives of former smelter workers in the service or recently deceased or disabled and with children or parents to support (Mercier 2001: 67–8). The conditions seemed clear; these women understood that they had to relinquish their positions to men at the war's end.

Although the company never advertised the smelter openings in late 1943, word spread quickly and many women eagerly applied. The opportunity to earn men's wages was a powerful lure. Ursula Jurcich, strapped to care for an invalid husband and young son, noted that, 'Everybody was talking about it, "oh, the women are working on the [smelter] Hill." That was a big "baloo" around here ... so I thought I might as well go and see if I can get on ... the money was big, that was important' (personal communication, Ursula Jurcich, August 1986). Nonetheless, the company's ambivalence and the firm community male breadwinner ideal led to just over a hundred women hired on the smelter (compared to almost 400 men over 60 years in age) during the war.

Wartime labour demands, and the subsequent relaxation of occupational barriers, presented unprecedented opportunities for women of many ethnicities in various industries, especially new aircraft and shipbuilding plants. Mining remained off limits however, except to a small percentage of women who found work in processing ore and on surface operations. Since 1890, Ontario mining legislation had prohibited the employment of women in mines. Calling on the *War Measures Act*, the Canadian government issued an order on 13 August 1942 that allowed women to be employed to allay a labour shortage, but only for surface operations. At International Nickel's (Inco) Sudbury operations, over 1400 women were hired for production and maintenance jobs during the war. They performed a variety of jobs such as operating ore distributors, repairing cell flotation equipment, piloting ore trains and working in the machine shop. At the end of the war, the government rescinded the order allowing the employment of women in the company's surface operations, and Inco saved the positions for returning servicemen (Sudol 2008).

Why did urban shipyards and aircraft plants actively recruit women, while mining communities resisted hiring them (see John 1980; Mark-Lawson and Witz 1988; Kingsolver 1989)? Practices were influenced by well-entrenched gender ideologies, accompanying lore about the work and past union struggles for job security in automating industries. The industry's needs and economic position also explain the difference; the minerals industry had been declining except during the war boom, whereas new industries like shipbuilding demanded new recruits. This helps explain why women never constituted more than five per cent of the smelter workforce in Anaconda, while they made up 28 per cent of Portland, Oregon's shipyard workers (Skold 1980: 57; Kesselman 1990).[2]

Even the left-leaning labour union, the IUMMSW, wavered from its expressed goal of gender equality given male rank-and-file resistance. In Anaconda, when the company tried to bypass union seniority rules to give women 'soft'

---

2   In her study of women war workers, Karen Anderson (1981) concludes that labour markets, rather than community values, determined such variations. She notes, for example, that women made up just four per cent of Baltimore shipbuilders, compared to 16 per cent in Seattle, because of greater availability of black male workers in the East.

positions, union leaders, struggling to preserve a male breadwinner workplace while maintaining labour principles, alternated between insisting that women be excluded as a weaker sex and that they be treated equally. By the end of the war, IUMMSW Local 117 sought to restore jobs to returning servicemen, but it also defended the women in a retroactive pay contractual issue. The union even sought to retain jobs for widows, who, as believed, merited continued employment because of their economic circumstances. But the union became caught in its own web of seniority rules, and the company insisted that the last woman hired would be the first to be laid off when a serviceman reclaimed his job. Within a few months no women were employed in production at the smelter.[3] Pearl Chytuk, who moved to Sudbury, Ontario, from Regina, Saskatchewan, in 1941, was able to get a job at the Inco smelter during the war, but was surprised that people were fearful of talking about unions. While working at the smelter, Chytuk actively organised for Mine Mill Local 598. She remembered the hesitancy of some of her male co-workers, but many of the women activists were, 'from the west where we always felt more free'. Despite their activism, the union could not help them retain their jobs at the war's end (Solski and Smaller 1985: 124; Steedman et al. 1995: 162–5).

The postwar period brought a rapid resumption of restrictions, and old gender barriers were re-erected as men returned home to their jobs and women were discharged, underscoring the fluidity of these ideologies and how pinned to power relations they were. Nonetheless, women in mining communities were often fundamentally changed by their wartime experiences, as represented by married women's increasing participation in the labour force. For example, women who worked in the Anaconda smelter during the war lost their jobs, but the independence they tasted lasted in subsequent work and family roles. Erma Bennett recalled that after the war people 'tried to change it back', but it was 'the beginning of the change' in women's roles as they sought greater public and economic participation (Mercier 2001: 91). In single industry mining communities, well-paying mining and smelting jobs remained enticing (if forbidden) to women until legislation made opportunities available again.

Labour Contraction in the 1970s

If women received a frosty reception in the mining workplace during the wartime increase in labour demand, their efforts to re-enter mines and smelters several decades later following government and union mandates presented even more challenges in the light of a declining industry. In the US and Canada,

---

3 For an extended discussion of the women smelter workers and gendered debates about employment during and after the war, see Mercier (2001: 64–77).

women won the legal right to enter male mining workplaces just as the industry began to mechanise, move operations out of the country and lay off workers in the 1970s. Federal affirmative action orders opened coal and hard-rock production jobs to women for the first time since the Second World War. For women in places like Arizona, Montana, Appalachia and Ontario, despite male resistance and harassment, the good wages offered made competition stiff for the few mining jobs available.

Despite initial opposition, female miners broke barriers in the UMW and created the Coal Employment Project (CEP) to combat discrimination, work on health and safety issues and form an international network of coalfield women. Although 1965 US Civil Rights legislation was amended in 1967 to include gender as a basis for non-discrimination, abolishing state prohibitions against women mining, social sanctions remained, as Carletta Savage (2000: 232) notes, 'in full force'. A class-action lawsuit settled in 1978 forced Appalachian coal companies to open their tunnels to women. In 1977 women made up just one per cent of the mining workforce; by 1979, they had increased to over 10 per cent (Moore 1996: xl–xlvi).

In the mid-1990s Suzanne Tallichet (2006) spent several months in a West Virginia coal mining community to uncover the world of women miners. In studying Bureau of Mines data the author found that gender was a more likely determinant of a miner's job rank than all other factors combined. In her research and interviews she determined that once hired, the 'daughters of the mountain' still had to negotiate tremendous barriers, including harassment from male co-workers and bosses and resistance from the community. Male miners asserted their solidarity and dominance by exaggerating gender differences and sexualising relationships with female co-workers. The interviewed women claimed that at least half of their co-workers and bosses targeted them with sexual and other harassment. To survive this treatment, more feared than the mine's dangers, women acted tough and repeatedly reminded the men of their practical needs: 'I'm not here for romance but for finance' (ibid.: 54). A woman had to prove her physical capabilities and accommodate men in various ways to win acceptance.[4]

But a 1981 CEP survey found that bosses, rather than co-workers, exacerbated women's problems underground. Tallichet (2006) and Savage's (2000) interviews with women miners confirmed that women were prevented from moving out of lower-paying strenuous jobs for more skilled positions. What was considered 'women's work' usually required the greatest physical endurance; supervisors rarely allowed women to learn new skills to operate machinery or they failed

---

4   Articles about women in mining frequently emphasise how the women had to show extraordinary stamina and good humour to survive male co-workers' bullying (see also, Sale 2007).

to train them when they did attain those positions, which caused resentment among male co-workers who were frustrated by the women's inexperience. A decade after women entered coal mines in large numbers, they still lacked training, endured high rates of sexual harassment and with little training and seniority were the first to be laid off when the industry declined in the 1980s and 1990s. Savage (2000) concludes that the very paucity of information about gender relations available to supervisors and personnel departments in the coal industry reflects how insincere they were about successfully integrating women into the workforce. The small monetary awards from harassment and discrimination lawsuits, and the hard labour and physical and emotional decline convinced many women miners to leave the industry altogether.

Resistance to women's employment did not emerge solely from male miners and employers. A surprising theme emerges from all of these studies, whether hard rock or coal, US or Canadian mining communities: the resistance from miners' wives was often the most virulent, upending any notions of gender solidarity. Tallichet (2006) found in her interviews that the wives viewed women miners as sexual and economic threats. Many wives even believed that women were not physically suited to working underground, endangering their husbands' lives. This called into question who was entitled to a breadwinning wage. In 1974, miners' wives in Logan, West Virginia, protested a local mine's hiring of women. As Marat Moore (1996: xxxvii–viii) discovered, male miners voiced greater support for affirmative action quotas than did their wives. Women miners blamed the wives for hardening men's attitudes towards them underground. There is a striking racial difference here, however. Although black women faced even greater resistance to mining employment, they experienced support, rather than hostility, from men and women in their community. These findings remind us that we need to question universal notions about women.

Diminishing opportunities were even more apparent for women who had edged their way into hard rock mining.[5] Jennifer Keck and Mary Powell (2006) outline how a hundred pioneering women took jobs at Inco in Sudbury, Ontario, in the mid-1970s after provincial legislation removed barriers to women's employment in surface jobs (1970) and later underground (1978). Motivated primarily by higher wages, women workers at Inco found that 'men's' jobs offered a great deal of satisfaction as well as greater financial independence. They endured the physically hard work, sexual harassment and difficulties finding child care to better support themselves and families. In adapting to masculine work culture, they had to prove they could 'do the work of a man' in order to be treated the 'same as a man'. In the process, they could achieve 'manhood' regardless of sex,

---

5   Although the coal industry also downsized and mechanised, reducing the numbers of workers, the oil and energy crises of the 1970s put more demand on production of domestic coal reserves, which sustained some demand for miners.

breadwinner status based on family need and respect for fighting back. During the strike of 1978–79 women workers saw themselves as workers first and declined to join wives in making sandwiches, instead joining their brothers on the picket line and in negotiating committees. Yet these challenges to the gender division of labour became muted as the minerals industry restructured and downsized in the 1980s; women and minority workers with the least seniority were laid off first.[6]

In the tight-knit copper smelting community of Anaconda, women remembered their Second World War predecessors, and with EEO openings in the 1970s seized opportunities for higher pay. But community attitudes towards the new generation of women smelter workers were different in a climate of retrenchment. As former wartime smelter worker, Katie Dewing (personal communication, August 1996) recalled, the women who crossed gender boundaries 'the second time around' in the 1970s 'had hard jobs and did all the things that men do' because they 'were taking men's jobs'. Yet despite workforce reductions, the Anaconda Coal Mine made an earnest effort to implement EEO hiring. Personnel director Bob Vine insisted that the women 'were readily accepted' by male workers. This may be because few women lasted. In March 1974 Carolyn Crisler left her position as a nurse's aid to work at the smelter. Since she weighed less than the required 130 pounds, she slipped weights into her pants to get the job because 'we were trying to buy a house, and we needed the money, and I knew if I got up there, I could save a lot of money.' Despite the hot, dirty and dangerous work in the converters, Crisler 'liked the crew … and the men accepted me and I did my work.' Nevertheless, she quit after two years because the physically exhausting work interfered with raising her young children. In 1980, the Atlantic Richfield Corporation, which absorbed the Montana ACM mines and smelters in 1977, shut down the state's operations.

The Arizona copper mining strike of 1983–85, involving the large multinational company, Phelps Dodge, and the Steelworkers Union represented a watershed moment in North American labour and mining history. Once again, women assumed a dominant role in maintaining the strike, loyally and passionately picketing for the same reasons they had in years past—to improve their families' living conditions, and to support striking men (and some women miners) who were legally enjoined from action. The strike ultimately failed and the company successfully decertified the union with its permanent replacement workers, ushering in a new non-union era in the American mining industry (Kingsolver 1989). Women miners and miners' wives in Arizona joined thousands of women

---

6 Gearhart (1998) notes that some women were displaced because of revived gender discrimination and the attraction of other work alternatives.

from other declining mining communities in North America to find other wage work, often in the expanding service sector, and often became the main breadwinners in their families.

Conclusion

As these examples from North America and scholarship from around the world demonstrate, women have persistently sought access to mining jobs. Women miners endured discrimination, harassment and dangerous and demanding labour in order to gain more comfortable lives for their families and in many cases the self-satisfaction that they could perform a 'man's' job. These cases remind us how masculine work cultures, entrenched corporate practices and social assumptions about gender, even in periods of labour expansion (such as the Second World War) or structural legislative changes (as in the EEO push of the 1970s), can impede the entry and survival of women in the mining workplace.

By the 1970s coal and minerals companies and the United Mine Workers and Steelworkers Union (successor to Mine Mill) officially opposed discrimination and harassment, but the oral sources often contradict official pronouncements. By the 1980s, many women had formed an uneasy alliance with their male co-workers in mines and smelters. But automation and industrial decline forced layoffs, and having the least seniority, women made up a larger proportion of those terminated. Although men and women together faced uncertain futures, women had spent a decade proving themselves, only to see men hang on to the few remaining skilled mining jobs.

In the twenty-first century, the mobility of capital, government policies, global structural adjustment programs and the quality and exhaustibility of ore continue to threaten the stability of mining communities. Despite its often attractive wages, mining continues to disrupt women's and indigenous agricultural practices, cultural traditions and extended kin and community networks; it displaces communities, introduces diseases and pollutes lands and water. Once led to extract better wages and working conditions from mining companies, recent protests often attempt to keep mining out of an area or challenge the ways in which multinational corporations extract resources. Tibetan protests against Chinese mining, Australian protests of uranium production and Honduran, Sri Lankan, and Indonesian protests of US corporate mining reveal the growing 'cultures of solidarity' between women, environmentalists, indigenous peoples, mining communities and international human rights organisations. Since the 1990s, several international conferences involving women in mining countries have met and an International Women and Mining Network has formed to protest the unsustainable nature of mining and its harm to women and communities (see, for example, IWMN: 2010).

As others have noted, women have not shared the benefits of mining, and have borne a disproportionate share of mining's negative effects. Excluded from mining or denied skilled training, women are denied higher wages, and in their more informal mining activities are more subject to mercury pollution and other health hazards. While we explore the ways that transnational corporations, national and local governments and international institutions can integrate women more fully in mining jobs, we might also ask whether mining—by its very nature socially and environmentally disruptive—can ultimately sustain women and men and their communities. Even with gender equity, can there be more positive mining developments?

Although these examples from North America may not seem compatible with the experiences of women in the developing world, I think we can see how gender ideologies take root and spread across time, space and cultures. Understanding this process can perhaps lead us to find ways to help move women from the borders to the center of mining workplaces and the decision-making processes that affect their lives.

References

Anderson, K., 1981. *Wartime Women: Sex Roles, Family Relations, and the Status of Women During World War II*. Westport: Greenwood.

Chaput, D., 1996. 'In Search of Silver and Gold.' *American History* 30(6): 36–40.

Finn, J.L., 1998. *Tracing the Veins: Of Copper, Culture, and Community from Butte to Chuquicamata*. Berkeley: University of California Press.

Forestell, N.M., 2006. '"And I Feel Like I'm Dying from Mining for Gold": Disability, Gender, and the Mining Community, 1920–1950.' *Labor: Studies in Working Class History of the Americas* 3(3): 77–93.

Francaviglia, R.V., 1998. 'In Her Image: Some Reflections on Gender and Power in the Mining Industry.' *Mining History Journal* 5: 118–26.

Gearhart, D.G., 1998. 'Coal Mining Women in the [American] West: The Realities of Difference in an Extreme Environment.' *Journal of the West* 37(1): 60–8.

Gier J. and L. Mercier (eds), 2006. *Mining Women: Gender in the Development of a Global Industry, 1670 to 2005*. New York: Palgrave Macmillan.

Gúerin-Gonzalez, C., 2006. 'From Ludlow to Camp Solidarity: Women, Men, and Cultures of Solidarity in U.S. Coal Communities, 1912–1990.' In J. Gier and L. Mercier (eds), op. cit.

IWMN (International Women and Mining Network (RIMM)), 2010. 'Campaigns'. Viewed 5 September 2010 at http://www.rimmrights.org/campaigns.htm

John, A.V., 1980. *By the Sweat of their Brow: Women Workers at Victorian Coal Mines*. London: Croom Helm.

Keck, J. and M. Powell, 2006. 'Women into Mining Jobs at Inco: Challenging the Gender Division of Labor.' In J. Gier and L. Mercier (eds), op. cit.

Kesselman, A., 1990. *Fleeting Opportunities: Women Shipyard Workers in Portland and Vancouver during World War II and Reconversion*. Albany: State University of New York Press.

Kingsolver, B., 1989. *Holding the Line: Women in the Great Arizona Mine Strike of 1983*. New York: ILR Press.

Klubock, T.M., 1998. *Contested Communities: Class, Gender, and Politics in Chile's El Teniente Copper Mine, 1904–1951*. Durham: Duke University Press.

Lahiri-Dutt, K. and M. Macintyre (eds), 2006. *Women Miners in Developing Countries: Pit Women and Others*. Aldershot: Ashgate.

Mark-Lawson, J. and A. Witz, 1988. 'From "Family Labor" to "Family Wage"? The Case of Women's Labor in Nineteenth-Century Coalmining.' *Social History* 13: 151–74.

Mercier, L., 2001. *Anaconda: Labor, Culture, and Community in Montana's Smelter City*. Chicago: University of Illinois Press.

MHSA (Montana Historical Society Archives), 1943. 'Plant Manpower Analysis.' 62/3-1943. Viewed 5 September 2010 at http://montanahistoricalsociety.org/

Moore, M., 1996. *Women in the Mines: Stories of Life and Work*. New York: Twayne Publishers.

Nash, J., 1979. *We Eat the Mines and the Mines Eat Us: Dependency and Exploitation in Bolivian Tin Mines*. New York: Columbia University Press.

Robbins, W.G., 1986. 'Labor in the Pacific Slope Timber Industry: A Twentieth-Century Perspective.' *Journal of the West* 25(2): 8–13.

Sale, A., 2007. 'Sisters in Coal: A History of Women in the West Virginia Mines.' *Goldenseal* 33(1): 10–7.

Savage, C., 2000. 'Re-gendering Coal: Female Miners and Male Supervisors.' *Appalachian Journal* 27(3): 232–48.

Schwantes, C.A., 1979. *Radical Heritage: Labor, Socialism, and Reform in Washington and British Columbia, 1885–1917.* Seattle: University of Washington Press.

———, 1982. 'Protest in a Promised Land: Unemployment, Disinheritance and the Origin of Labor Militancy in the Pacific Northwest, 1885–86.' *Western Historical Quarterly* 13: 373–90.

———, 1988. 'Images of the Wageworkers' Frontier.' *Montana: The Magazine of Western History* 38: 38–49.

Skold, K.B., 1980. 'The Job He Left Behind: American Women in the Shipyards During World War II.' In C.R. Berkin and C.M. Lovett (eds). *Women, War, and Revolution.* New York: Holmes & Meier.

Solski, M. and J. Smaller, 1985. *Mine Mill: The History of the IUMMSW in Canada since 1895.* Ottawa: Steel Rail Publishing.

Steedman, M., P. Suschnigg and D.K. Buse (eds), 1995. *Hard Lessons: The Mine Mill Union in the Canadian Labour Movement.* Toronto: Dundurn Press.

Sudol, S., 2008. 'Inco's Sudbury Nickel Mines were Critical during World War Two.' *Republic of Mining.* Viewed 4 October 2008 at http://www.republicofmining.com/category/women-in-mining/

Tallichet, S.E., 2006. *Daughters of the Mountain: Women Coal Miners in Central Appalachia.* University Park, Pennsylvania: Pennsylvania State University Press.

Yao, L., 2006. 'Women in the Mining Industry of Contemporary China.' In Kuntala Lahiri-Dutt and M. Mcintyre (eds), op. cit.

Zanjani, S., 1997. *A Mine of her Own: Women Prospectors in the American West, 1850-1950.* Lincoln: University of Nebraska Press.

4. Sex Work and Livelihoods: Beyond the 'Negative Impacts on Women' in Indonesian Mining

Petra Mahy[1]

Introduction

The highly gendered nature of the global mining industry has now been exposed by a number of academic commentators and activists who advocate the rights of local communities in relation to mining companies. This research is also now being recognised by international bodies such as the World Bank and the International Council on Mining and Metals (ICMM).[2] Large-scale mining around the world provides overwhelmingly male-dominated employment and tends to create and reinforce masculine-oriented cultures in the workplace and in mining towns. It is often argued that, as a consequence of this, women have been excluded from the direct economic benefits of mining and have borne the brunt of any negative social and economic changes. While the evidence for these gender imbalances is clear, I argue here that the prevailing emphasis on the 'impacts' of mining on women creates and reinforces problematic categorisations and the essentialising of women living in mining communities, particularly in reference to commercial sex.

The almost inevitable growth of the sex industry in mining towns has been written about by both academics and activists. Some scholars investigating social change in mining areas have described in detail the masculine cultures that promote sex industries in mining towns (Robinson 1996; Campbell 1997), and a small number of researchers have looked at the issue from sex workers' points of view (Campbell 2000; Kunanayagam 2003). In addition, Parpart (1988,

---

1   The research for this chapter was supported by the Australian Research Council Linkage Project with PT Kaltim Prima Coal (KPC). I acknowledge the assistance of KPC staff, especially Yuliana Datubua and Nurul Karim, and also thank Kuntala Lahiri-Dutt for her comments on this chapter.
2   The World Bank has recently established an Extractive Industries and Gender Initiative after sponsoring a series of women and mining conferences in Papua New Guinea. The Mining, Minerals and Sustainable Development (MMSD) project's 2002 final report 'Breaking New Ground' and some subsequent ICMM output documents contain cursory recognition of gendered differences in mining impacts. More recently, the World Bank established an Extractive Industries and Gender Initiative after sponsoring a series of women and mining conferences in Papua New Guinea.

2001) has documented patriarchal attempts to restrain the increase in 'wicked women' on the Zambian Copperbelt as well as women's resistance to these policies of control.

However, the majority of publications on mining and sex work are by 'women and mining' activists. In these writings, prostitution is simply listed among other 'impacts on women' as though it is self-evidently negative (for example, see Bhanumathi 2002; Macdonald 2002, 2006; Byford 2003; Hill 2007, 2008; Eftemie 2008; Oxfam Australia 2009: 7). While undeniably negative consequences can certainly be traced or predicted, such as an increase in HIV infection,[3] I argue that this activist literature presents an overly simplistic picture of mining community women's actual diverse relationships to the sex industry, whether as sex workers, miner's wives or as women otherwise living in a mining community. It also overlooks the capacity of women and men to protest and to act.

By relying on moral reactions to prostitution and neglecting to explain exactly why it is that prostitution in mining areas is negative for women, this literature implies two alternative, yet equally problematic, categorisations of women living in mining towns.

The first alternative assumes that women, as one group, regardless of occupation or choice, are equally the victims of mining and its resultant market for commercial sex. This approach either entirely neglects inclusion of sex workers in the category of 'mining community women', or it includes them but assumes that selling sex is an inherent violation of women's human rights regardless of a sex worker's choice or agency.

The second alternative distinctly divides women living in mining communities into two categories: (migrant) sex workers and (indigenous) community women. Community women are assumed to be the chaste, ignorant and powerless victims of their husbands' sexuality and their assumed predilection for spending their mining wages on the services of prostitutes. If an indigenous community woman should become a sex worker then she is seen as having been forced by social and economic circumstances resulting from mining into a demeaning occupation. On the other hand, migrant sex workers must necessarily be opportunistic 'bad' women who take advantage of 'good' women's husbands and the blameworthy vectors of disease rather than being identified as individual women pursuing livelihood options. This second alternative brings the forced (innocent) or voluntary (guilty) dichotomy into the debate about sex work in mining towns.

---

3   Many large-scale mining companies now have HIV/AIDS prevention programs of one sort or another, having recognised the health threat to workforces and local communities. Female sex workers are often the targets for these programs as a 'high risk population' that has direct links back to the male dominated mining labour force.

While the literature which implicitly relies on these two alternative categorisations may help to motivate change in mining policies, it does not accord with many feminists' rejection of the notion of women necessarily being 'victims' without agency. Nor does it accord with new understandings of sex work. Nor, further, is it reflective of the realities in the mining community where I conducted my research. This chapter aims to move beyond these problematic categorisations and to present a more balanced portrayal of the sex industry in a mining area. It is based on research carried out during 2007 among the communities surrounding Kaltim Prima Coal (KPC), a large-scale coal mine in Indonesia. This research includes in-depth interviews with female sex workers and others involved in the sex industry as well as with many women and men otherwise living in the mining area from diverse economic and social backgrounds. While every mine is necessarily different and situated within a specific temporal, social and economic context, that this one mining area so evidently does not fit in the straightforward 'all women as victims of mining' model, it at least suggests that accounts of gender and sex work in other mine areas should be re-examined.

Victims or Heroines, Prostitutes or Sex Workers?

At the heart of this issue of causally linking mining with 'negative impacts on women' is the fundamental feminist dilemma of the tendency to portray women as being essentially similar in their 'victimhood' in relation to patriarchy or patriarchal development programs in order to motivate cohesive action. The (often postcolonial) representation of the essential sameness of women as an 'already constituted and coherent group with identical interests and desires, regardless of class, ethnic or racial location' (Mohanty 1988: 64), has long been shown to be misleading by overlooking the diverse lived experiences of women and thereby potentially increasing disempowerment and stigma. The general 'women and mining' literature likewise tends to assume that all women living in mining areas have similar backgrounds and interests (and particularly are all indigenous, rather than migrants), and that this similarity existed prior to and after the start of large-scale mining (see Macdonald 2002, 2006).

Beyond this one essential category, women also often feature in these narratives in dichotomous identities, for example, as either victims or heroines (Cornwall et al. 2007). They are the victims of imposed economic development programs that interrupt their supposedly closer natural connection to the land and are afflicted by male oppression, alcoholism, violence and sexual urges. The heroines are the women bearing these burdens, yet struggling bravely and opposing development. While these victim/heroine narratives have the power to initiate action and policy change (as evidenced by the women and mining campaigners' success in gaining the attention of the World Bank), they fail to truly reflect gendered realities.

Singular and dichotomous representations of women also run through the debates about prostitution/sex work and the trafficking/migration of women for sex. Feminist analysis has shown that these debates have a tendency to fall into the trap of innocent/guilty and madonna/whore representations.

Some, often termed abolitionists, see prostitution as an extreme form of gender discrimination that is 'inherently violative of women's bodily integrity and freedom from violation, regardless of consent or choice' (Peach 2008: 237). Abolitionists see female prostitutes as needing to be saved from commoditisation and a life of sexual slavery (for example, Barry 1995). On the opposing side of the prostitution and anti-trafficking debate are those 'reformists' and sex worker activists who argue that prostitution should be seen as sex work—a legitimate form of labour, and not an inherently evil or immoral practice. Acknowledging that some women are indeed forced into selling sex and their human rights violated, the reformists (and UN anti-trafficking instruments) draw a distinction between forced and voluntary sex work. Critics of this approach have in turn argued that the forced/voluntary dichotomy creates a guilty/innocent division that reproduces the whore/madonna division within the category of the prostitute (Doezema 1998: 47). The forced/voluntary approach has been shown not to reflect the reality of sex workers' multiple subjectivities and personal agency within the context of wider social, economic and personal factors (Kempadoo 1998; Sandy 2006).

It has also been argued that due to all the attention on trafficking, women who migrate for sex have long been missing from migration studies thus preventing them from being seen as members of diasporas, as entrepreneurial women and active agents participating in globalisation (Agustín 2006, 2007). Granting agency to individuals who migrate for sex work does not mean denying the vast structural pressures that push and pull them (Agustín 2006: 39), rather it allows acknowledgement that migration for the sex industry is often a way of 'expanding life choices and livelihood strategies' (Doezema 2000: 26). It can also be a means of travel and seeking adventure (Bandyopadhyay et al. 2007: 95). These arguments are supported by a study in the Riau Islands, Indonesia, which describes how becoming a migrant sex worker allows some women to make a living, to provide for family members and, in some cases, to find their way into a more prosperous lifestyle (Ford and Lyons 2008).

My research in the KPC area shows that this line of feminist research on migration and sex work that rejects the forced/voluntary division should be incorporated into discussions of the gendered dynamics at work in mining communities.

Female sex workers in mining areas should be acknowledged as women who are pursuing a livelihood opportunity within their wider socio-economic context. Sex workers are not necessarily victims or heroines, nor are the women living in mining communities who pursue other means of economic support.

Selling Sex in a Mining Area: KPC Research Findings

One of the largest coal mines in Indonesia, Kaltim Prima Coal (KPC) began work in the early 1990s and has a contract of work until 2021. It was owned equally by Rio Tinto and BP until 2003 when the company was sold to the Indonesian company Bumi Resources. A minority share has recently been bought up by the Indian company Tata Power. KPC operates in two main areas: Sangatta and Bengalon, in the district of East Kutai in East Kalimantan, Indonesia. Mine production began in Sangatta in 1991 and in Bengalon in 2004. These areas were sparsely populated with indigenous and migrant farmers prior to KPC's operations, but have now attracted large numbers of migrants from around Indonesia seeking employment and business opportunities in the area.

Employment statistics show that around 95 per cent of workers in KPC are men. KPC's contractor companies tend to have an even higher proportion of male employees. At KPC, the 5 per cent of women workers tend to be clustered in administrative roles, although there are women geologists, engineers and truck operators. The proportion of women employees has decreased slightly over time from a high of about 7.5 per cent in 1993 (Lahiri-Dutt 2004) to 5 per cent in 2006. Some women informants explained that it was easier for the wives of company employees to obtain work in the company in the earlier years of mine production when there was a smaller local labour supply.

Demographic data shows consistently higher numbers of men moving to the area than women, particularly far higher numbers of single men than single women.[4] My interview data shows that an overwhelming majority of married women cite their reason for migrating to the KPC areas as, following their husbands (*ikut suami*), although this answer may well gloss over their role in the migration decision. It is necessary, however, to look beyond this majority and recognise that there is plenty of evidence of single women migrating there by choice to live with more distant relatives, to find well-paid husbands or to find work, some as sex workers. There is also ample evidence of single and married women earning incomes in entrepreneurial roles, shops, farming, government work

---

4   These statistics are derived from the Indonesian Population Census of 2000 and the Indonesian Bureau of Statistics Kutai Timur dalam Angka [East Kutai in Numbers] series.

and as teachers, among other roles. Female migrants in my surveys consistently reported a higher standard of living in the KPC areas than in their place of origin and particularly appreciated the ability to more easily finance their children's education in the mining area (Lahiri-Dutt and Mahy 2006).

Class is also a significant factor in women's lives in the KPC areas. The lives of the women who live in the higher-level mine accommodation compounds or the family of top government officials are significantly different to the lives of women in farming, petty trade or mining operators' families. Ethnicity and religion play a part in determining where certain groups are more likely to live and the social circles they move in but do not appear to be as important a determinant as class to lifestyles and status in the multi-ethnic communities around KPC.

Sex Industry: Changes over Time

Prostitution is not just a single or static practice and the sex industry in KPC areas of operation has changed in character over time. This change has taken place according to the mine cycle and also in response to changes in the legal and illegal logging industries. It is generally agreed that the sex business was at its busiest during the mine construction phases when mine employment was dominated by single men working on short contracts. In the early 1990s, when the mine workforce included many male expatriates, there was a system of 'contract wives' in place under which bar girls would each try to contract more regular services and payment with an expatriate worker. These arrangements would often create the illusion of romance with some, sometimes going as far as holding contract wedding ceremonies (Kunanayagam 1994).

It is also reported that KPC was notorious as one of the easiest places for workers to institute such arrangements, and that women were often allowed to live inside the housing compounds for long stretches of time; they were also given their own security passes. It is said that at KPC 'everyone had a girl in their room' (Cannon 2002: 214). This practice has now been prohibited by stricter security measures and tighter allocation of catering facilities. The drop in numbers of expatriates has also had a significant effect, and the sex industry in the KPC areas is now dominated by Indonesian clientele. The contract wife system has now mostly disappeared and many bars have gone out of business or moved away from areas that are now more family-oriented.

Currently, the main area where sex is sold in Sangatta is the quasi-official brothel complex (*lokalisasi*) (Kampung Kajang), with around 150 resident sex workers occupying a number of different brothels (*wisma*) facing a narrow lane. Each *wisma* also has a bar which serves beer and provides music and karaoke entertainment. The complex has a male coordinator who manages the young male security guards and other matters. A 2005 survey estimated that approximately 20 per cent of clients to Kampung Kajang are male mine workers; the remainder includes civil servants, army and private sector workers and students (*Tribun Kaltim*, 2005). There are also numerous separate bars, small hotels, massage parlours and cafes (*warung*) located elsewhere in Sangatta where sexual services are sold. Some of these are strategically located in relation to mine worker accommodation.

In Bengalon, there is a similar sized *lokalisasi* located on an unsealed road outside the main town. It is said that there has been prostitution in the Bengalon area since around 1986 with mainly logging clients, but the busiest years in Segadur were 2000 and 2001. Closer to the mine and coal port area there is a number of bars. Again, during the mine construction period here, there was more business than there is currently with more short-term contract workers present. Two of the bars in this area used to operate in Segadur, but have now moved in order to be closer to the mine in the hope of better business and less competition. At the time of data collection (late 2007) there were 20 female sex workers working in the three bars in this area.

Both *lokalisasi* and bars in Sangatta and Bengalon have obviously been affected by the government crackdown on the illegal logging industry. Most sex workers and brothel owners commented that business had been comparatively quiet in the previous year or so.

A factor common to the *lokalisasi* and bars in these areas is their integration in to local economies. Many people depend vicariously on them for their livelihoods. These include cafe owners, masseuses, parking attendants, laundry workers, security guards, clothing and cosmetics peddlers and motorcycle taxi drivers. Profits also flow to the police and local government in the form of administrative fees.

Travelling to East Kalimantan to Sell Sex

Most sex workers in the Sangatta and Bengalon *lokalisasi* and bars are of East Javanese and Madurese origins. Other ethnic identities found in smaller numbers include Sundanese, Buginese and Dayak. None of the sex workers interviewed claimed to have been born or grown up in the mining areas, and, like the majority of adult residents in the area, can all be classed as migrants. Their ages range from teenage to above 40 years, with most of the women in their 20s and 30s. Most of the sex workers come from poor rural or urban backgrounds and have low levels of education, though a small number had completed high school.

A majority of the sex workers interviewed were divorced or separated with children whom they support through their work in East Kalimantan. Their children tend to live with grandparents in their home towns. A small number of the sex workers had never married or were single at the time. The women from East Java and Madura in particular describe a pattern of early arranged marriages that did not last very long. Many share stories of desertion—their husbands remarrying, of husbands' inability to support them or simply, incompatibility. One woman was escaping her husband's creditors. She was slowly paying off his debts by working as an assistant brothel manager and bar tender. A twenty-year-old woman had run away from home in Semarang in order to escape an arranged marriage. Another had brought her baby to Sangatta with her but fostered the child with a local farming family.

Many of the sex workers do not tell their families about the kind of work they do; instead they inform them that they work in shops, hotels, as domestic servants or run small businesses. Some families, however, particularly those who collect remittances at the brothel manager's home in Java, are well aware of the type of work that the women do in Kalimantan even if it is not spoken about openly. A few interview extracts have been provided:

> C: Some of us have just passed primary school or junior high school. Our parents did not have the money to send us to school. We are trying to earn our own money so we don't need to ask our parents for it, so that our children will not become like us.

> L: I'm from Malang in East Java. I'm divorced. I followed a friend here. I heard about Kalimantan from that friend. Java is far more developed, but it is easier to earn money here. I have two children. My ex-husband's parents look after them. They love my kids like I would, maybe more.... Every month I send home money for my children, but the amount I send depends on the number of clients I get. Sometimes it is very quiet. I would like to work somewhere else one day, become a good person, but it is not yet possible.

While most cited their economic circumstances and lack of a male breadwinning partner as reasons for their decision to enter the sex industry, a small number of the female sex workers placed more emphasis on the fact that this lifestyle was less confining than being married and obeying one's husband:

> V: I understood what it meant to be a housewife, but my husband was authoritarian. I was young and wanted to sleep in until the afternoon every day. If I wanted to go somewhere I had to ask for his permission. I was stressed and depressed. Now I'm divorced. On the other hand, now

I'm free. If I compare now to when I had a husband, it is better now....
I don't call my family as it would just make them sad. I can't tell them
that I work like this, right.

While many sex workers feel isolated and sad about their circumstances,
others quite enjoy their rebellious status and breaking mainstream norms for
Indonesian women. These women flaunt their ability to wear sexy clothing, to
dance until late at night, drink alcohol, smoke cigarettes and have tattoos. 'I
got it done when my husband left me,' said one woman about the rose tattoo on
her shoulder. Most of the sex workers used assumed names either to hide their
original identities or to assume a more sophisticated persona.

The sex workers most often spoke about their situation with the euphemism
'working like this' (*kerja begini*), while some would use the term 'PSK'—
the Indonesian acronym for a commercial sex worker. Invariably they called
themselves 'workers'. Others in the community would use PSK or '*wanita nakal*'
(naughty/bad women) to refer to them. A small number of the sex workers
described themselves as business women:

> D: I want to earn my own money. To save money. After all there's no
> guarantee that I'll find a rich husband. When I was first here I was
> earning a lot of money. While my friends were sending money home
> I was saving. Now I'm not earning quite as much. But I've been able
> to buy a house, and motorbikes for all my younger siblings, one by
> one, and started some transport businesses.... I'm good at business. I've
> studied what my rich relatives in Jakarta do and copied them.

High Mobility in the Sex Industry

There are many areas where sex is sold in East Kalimantan: in urban centres,
transport hubs and in rural districts where there is industry of some kind to
support it. In the KPC areas, the sex workers had mostly come through informal
recruiting networks. One common way is for current sex workers to go home
for holidays and then return to East Kalimantan with friends who are seeking
work. Ramadan and Idul Fitri are common times for the women to return home
as business tends to be slower during the fasting month and it is the traditional
time for migrant workers to visit their families. Often the cost of an airfare to
East Kalimantan is borrowed from the brothel owner, and the sex worker can
return only having cleared the debt. Some of the sex workers can fall into cycles
of debt with the brothel owner. Other networks are more formal; brothel owners
often recruit in their home regions and maintain a house there where women
seeking work can contact them:

Y: I didn't come with a friend. I came by myself. All because I was annoyed with my husband who went to Malaysia and didn't come back for five years. I found out he had married again. So I ran away. That's all. I work. My brother-in-law helped take me to a particular house and then I came here. My mother cried when I left. I've been home once and then I came back here. My father is old and cannot work. I'll stay here while my child is at school. I don't care if my husband brings his new wife home—the important thing is that my child isn't taken away.

Some of the women claimed that they were tricked into coming and expected some other form of employment. On finding that they were expected to be sex workers they accepted the change for lack of better choices. There was one documented case of trafficking in 2006 where a brothel owner in Kampung Kajang was imprisoned for 18 months for receiving a young woman who had been brought to Kalimantan with the expectation that she would work in a shop and then reported her situation to the police. That this one woman at least was able to protest suggests that the others who were tricked but stayed had made conscious decisions to do so.

There is a high level of mobility among the sex workers. Many, though not all, had worked elsewhere in East Kalimantan before coming to Sangatta or Bengalon, including complexes in Samarinda and Tarakan. While some stay for a number of years, others stay for only a week or two if they dislike the place and prefer to seek better opportunities elsewhere. Some have moved between Kampung Kajang and Segadur and other bars in the area:

T: I have been here for three years. It is better here than when I worked in Tarakan for eight months. It wasn't a city area. There were little rivers everywhere and they were the only places we had for bathing. If there were any disturbances then we had to run away ourselves, find our own safety. There was no security…. Here, I used to be contracted to one man [an illegal logger], but it is better to find a client every night because that way I'm free.

A: My parents arranged a marriage for me when I was 12 years old. Finally I ended up here. I didn't know about this area though I'd been past often. I just followed a friend. I'm half Kutai and half Dayak Kenyah. First I worked in Segadur for three weeks, then I came here. In Segadur there were lots of wismas, there was a lot of competition. Here the Mami [female brothel owner] is nice and so is the food. In Segadur the music has to be turned off at 12am. Here we can play music until morning. There my earnings disappeared on electricity bills and food….Actually,

I can't save money, it just slips through my fingers. I buy clothes. I give it to friends if I'm feeling sorry for them ... before it was ... well now I'm independent. I can travel far and have my own income.

Women who leave the complexes either return home, get married and live in the community, or move to other *lokalisasi* or bars within East Kalimantan. Bar owners are often former sex workers. Those who stay on in the community are often whispered about as being ex-prostitutes and they may try to hide that part of their lives as far as possible. I found that there were different levels of tolerance towards former sex workers depending on the social circles to which they belonged.

Choosing Work Conditions

The sex workers in the KPC area demonstrated that they were weighing up their options and making choices about where they would work, particularly whether they preferred to live in the *lokalisasi* or in individual bars. They make decisions based on the cut of their earnings that they must give their *wisma* owner, whether they must pay for electricity and water use, and whether they are required to buy their food from the *wisma* owner or if they are free to eat anywhere. Unlike the bars, sex workers in the *lokalisasi* must pay weekly and monthly fees to the complex coordinators to fund the security officers and to pay police and local government administration fees.

The advantage of the *lokalisasi* is that it clearly provides a reliable security system for the sex workers. Should one of the women have any trouble with a client due to drunkenness, violence or, more commonly, a refusal to pay, she shouts for help and the security officers will chase down the man and demand payment. In Kampung Kajang, truncheons are kept ready in the security booth. No one can enter or leave the complexes without passing it. None of the sex workers interviewed reported any cases of personal violence against them that the security officers could not handle. While it is possible to leave the complex at night with a client by paying an extra fee, most say that they would not do so unless the client was a trusted regular. The individual bars tend not to provide this kind of security, and in some the workers routinely leave with the client.

The *lokalisasi* security system does, however, represent a check on the sex workers' freedom of movement. In order to leave during the day, to go shopping or to the bank or the doctor, they must first seek permission from their *wisma* owner and then report to the security post and pay a small fee before leaving. They are also usually required to use motorcycle taxis (*ojek*) that are affiliated with the complex. The security officers also check that the sex workers are dressed appropriately for appearing in public. The complex coordinators say that this system is required in case 'anything should happen' to the women while away. On one occasion, I was present when two sex workers had returned to the *lokalisasi* after having left without getting permission or paying the fee. They were being sternly spoken to by the complex manager, but shrugged and laughed afterwards. They had decided that it was worth risking the punishment to avoid paying the fee.

Agency from Different Perspectives

The district government has periodically announced plans to shift the *lokalisasi* at Kampung Kajang to a more remote location in order to remove 'immoral practices' from the general population (*Kaltim Post*, 2007a, 2007b, 2007c, 2007d). No concrete steps have been taken as yet and where the budget would come from for the move is contentious. The coordinator of the complex pointed out that any relocation would be far more difficult than the government has bargained for. There would need to be buildings in the new location and a serviceable road and it would disrupt the livelihoods of not only the *wisma* owners and sex workers but all the other people who rely on the complex for their incomes. 'The *lokalisasi* was here first, and all these others moved in to live around us, so trying to remove the complex from the general population is absurd,' he said.

The economic dependency of many women on their husbands or other male relatives is a reality in the KPC areas given the larger labour market for men and mainstream norms about the nuclear family and child care. While one can find plenty of harmonious households with apparently high levels of trust, there are also those where wives expressed their concern that their husbands are cheating on them or spending wages on alcohol, gambling and/or sex. One woman commented, 'here a husband might say he's going out to a coffee shop and come back a father.' In some cases increased income has prompted men to take on second wives or mistresses.

However, it is wrong to think that women are necessarily passive in the face of such pressures. In 2006, a group of women in a Bengalon village narrated how they had came together as a group to successfully close down a nearby bar in response to their fears that the sex workers there would tempt their husbands away. In fact one village man had left his family in order to marry one of the sex workers. The women said that they were also worried about the young men from the village spending their wages on alcohol at the bar. This bar in fact did not close entirely but moved somewhat further away. The bar owner in question when I asked him about this, agreed that he had been forced to move. He also admitted that the profits from his bar depend on having female waitresses and sex workers.

Nor will many wives passively allow their husbands to visit sex workers. In the *lokalisasi*, I was told that it is not uncommon for a wife to seek her husband in the brothel complexes and to take him home. The complex management does not interfere so long as there is no violence or disruption. They may even extract a man from the complex and deliver him to his wife if she waits outside the complex. One informant, a business owner, admitted that she had had some trouble with her husband being tempted by women and gambling in the past, but that she had made her feelings clear on the matter and he had changed his behaviour. It helped in her case that she was the main income earner in the family and could determine how much money her husband had access to.

Conclusion

Sex work in mining towns is a vexed issue with plenty of varying perspectives depending on one's moral and/or feminist viewpoints. This research shows that while negative consequences certainly exist, sex work should not be straightforwardly classified as a negative impact of mining on *all* women. In the absence of better choices, it often provides a livelihood and an escape from mainstream social constrictions for many women. The sex workers in the KPC area were clearly expressing the reasoning behind the choices they have made both in becoming a sex worker and concerning the conditions of their work. The victim/heroine and forced/voluntary dichotomies have very little meaning in this context.

Where sex workers' agency is acknowledged, so too is the capacity of miner's wives to protest and to act in ways that ensure their own economic survival and personal satisfaction. The picture of the chaste, ignorant and passive woman living in a mining community needs to be replaced with a far more real picture of diversity, opportunism and agency while also acknowledging that mining communities have specific gendered dynamics caused by the male dominated workplace and wider patriarchal traditions.

The 'women and mining' literature has been moderately successful in bringing to light the previously overlooked gender disparities in mining communities. However, policy makers should be cautious when implementing any changes based around depictions of homogenous categories of women in mining communities. In the context of sex work, this is particularly relevant for HIV prevention programs and the targeting of high risk populations.

References

Agustín, L., 2006. 'The Disappearing of a Migration Category: Migrants who Sell Sex.' *Journal of Ethnic and Migration Studies* 32(1): 29–47.

———, 2007. *Sex at the Margins: Migration, Labour Markets and the Rescue Industry*. London and New York: Zed Books.

Bandyopadhyay, N., S. Gayen, R. Debnath, K. Bose, S. Das, G. Das, M. Das, M. Biswas, P. Sarkar, P. Singh, R. Bibi, R. Mitra and S. Biswas, 2007. '"Streetwalkers Show the Way": Reframing the Debate on Trafficking from Sex Workers' Perspective.' In A. Cornwall, E. Harrison and A. Whitehead (eds), *Feminisms in Development: Contradictions, Contestations and Challenges*. London: Zed Books.

Barry, K., 1995. *The Prostitution of Sexuality*. New York: New York University Press.

Bhanumathi, K., 2002. 'Mines, Minerals and PEOPLE, India.' In I. Macdonald and C. Rowland (eds), op. cit.

Byford, J., 2002. 'One Day Rich: Community Perceptions of the Impact of the Placer Dome Gold Mine, Misima Island, Papua New Guinea.' In I. Macdonald and C. Rowland (eds), op. cit.

———, 2003. 'Too Little Too Late' Women's Participation in the Misima Mining Project.' Paper presented at conference on 'Women in Mining: Voices for Change', Madang, 3–6 August.

Campbell, C., 1997. 'Migrancy, Masculine Identities and AIDS: The Psychosocial Context of HIV Transmission on the South African Gold Mines.' *Social Science and Medicine* 45(2): 273–81.

———, 2000. 'Selling Sex in the Time of AIDS: The Psycho-social Context of Condom Use by Sex Workers on a Southern African Mine.' *Social Science and Medicine* 50(4): 479–94.

Cannon, J., 2002. Men at Work: Expatriation in the International Mining Industry. Melbourne: Monash University (Ph.D. thesis).

Cornwall, A., E. Harrison and A. Whitehead, 2007. 'Gender Myths and Feminist Fables: The Struggle for Interpretive Power in Gender and Development.' *Development and Change* 38(1): 1–20.

Doezema, J., 1998. 'Forced to Choose: Beyond the Voluntary vs. Forced Prostitution Dichotomy.' In K. Kempadoo and J. Doezema (eds), op.cit.

————, 2000. 'Loose Women or Lost Women? The Reemergence of the Myth of White Slavery in Contemporary Discourses in Trafficking in Women.' *Gender Issues* 18(1): 23–50.

Ford, M. and L. Lyons, 2008. 'Making the Best of What You've Got: Sex work and Class Mobility in the Riau Islands.' In M. Ford and L. Parker (eds), *Women and Work in Indonesia*. New York: Routledge.

Hill, C., 2007. 'Reframing the Mining Debate: Demystifying Paradigms and Mobilising Global Resistance. Oxfam Australia's Approach.' Paper presented at 'Inter-disciplinary Conference on Mining in the Asia-Pacific', Philippines, 26-8 November.

————, 2008. 'Women and Mining: The Role of Gender Analysis.' Paper presented at workshop on Mining, Gender and Sustainable Livelihoods, Canberra, 6–7 November.

Kaltim Post, 2007a. 'Awang, Lokalisasi Harus Ditertibkan! Pemerintah tak akan Biayai Relokasi ke Kampung Kajang.' [Awang Announces Prostitution Complex must be put in Order! The Government will not Pay for the Relocation of Kampung Kajang.] 10 January.

————, 2007b. 'Relokasi Lokalisasi.' [Relocation of Prostitution Complex.] 12 January.

————, 2007c. 'Tempat Pelacuran Belum Pindah: Bupati Beri Deadline Juni.' [Prostitution Area has not Moved: Bupati gives June Deadline] 4 June.

————, 2007d. 'Kampung Kajang Ditutup.' [Kampung Kajang to be Closed.] 8 November.

Kempadoo, K., 1998. 'Introduction: Globalizing Sex Workers' Rights.' In K. Kempadoo and J. Doezema (eds), op.cit.

K. Kempadoo and J. Doezema (eds), 1998. *Global Sex Workers: Rights, Resistance, and Redefinition*. New York: Routledge.

Kunanayagam, R., 1994. 'In the Shadow of the Company: Patronage and Status in a Mining Settlement.' Melbourne: Monash University (Masters thesis).

————, 2003. 'Sex Workers: Their Impact on and Interaction with the Mining Company.' Paper presented at conference on 'Women in Mining: Voices for Change', Madang, 3–6 August.

Lahiri-Dutt, K., 2004. 'Gender Survey of Kaltim Prima Coal.' Unpublished report to Kaltim Prima Coal.

Lahiri-Dutt, K. and P. Mahy, 2006. 'Impacts of Mining on Women and Youth in Indonesia: Two Mining Locations', Unpublished report to World Bank. Viewed 17 July 2010 at http://empoweringcommunities.anu.edu.au/documents/MiningImpactsReport.pdf

Macdonald, I., 2002. 'Introduction: Women's Rights Undermined.' In I. Macdonald and C. Rowland (eds), op. cit.

———, 2006. 'Women Miners, Human Rights and Poverty.' In K. Lahiri-Dutt and M. MacIntyre (eds), *Women Miners in Developing Countries: Pit Women and Others*. Aldershot: Ashgate.

Macdonald, I. and C. Rowland (eds), 2006. *Tunnel Vision: Women, Mining and Communities*. Fitzroy: Oxfam Community Aid Abroad.

MMSD (Mining, Minerals and Sustainable Development Project), 2002. *Breaking New Ground - Mining, Minerals and Sustainable Development*. London: Earthscan.

Mohanty, C., 1988. 'Under Western Eyes: Feminist Scholarship and Colonial Discourses.' *Feminist Review* 30: 61–88.

Oxfam Australia, 2009. *Women, Communities and Mining: The Gender Impacts of Mining and the Role of Gender Impact Assessment*. Carlton: Oxfam Australia.

Parpart, J.L. 1988. 'Sexuality and Power on the Zambian Copperbelt: 1926–1964.' In S.B. Strichter and J.L. Parpart (eds), *Patriarchy and Class: African Women in the Home and the Workforce*. Boulder and London: Westview Press.

———, J.L. 2001. '"Wicked Women" and "Respectable Ladies": Reconfiguring Gender on the Zambian Copperbelt, 1938-1964.' In D.L. Hodgson and S.A. McCurdy (eds), *"Wicked" Women and the Reconfiguration of Gender in Africa*. Portsmouth: Heinemann.

Peach, L.J., 2008. 'Female Sexual Slavery or Just Women's Work? Prostitution and Female Subjectivity within Anti-trafficking Discourses.' In K.E. Ferguson and M. Mironesco (eds), *Gender and Globalization in Asia and the Pacific: Methods, Practice, Theory*. Honolulu: University of Hawai'i Press.

Robinson, K., 1996. 'Women, Mining and Development.' In R. Howitt, J. Connell and P. Hirsch (eds), *Resources, Nations and Indigenous Peoples: Case Studies from Australasia, Melanesia and Southeast Asia*. Oxford: Oxford University Press, Oxford.

Sandy, L., 2006. 'Sex Work in Cambodia: Beyond the Voluntary/Forced Dichotomy.' *Asia and Pacific Migration Journal* 15 (4): 449–69.

Tribun Kaltim, 2005. 'Satu PSK Kutim Mengidap HIV: Pelajar Juga Sering ke Lokalisasi.' [One East Kutai Sex Worker Infected with HIV: Students also often Visit Prostitution Complexes.] 31 August.

5. Experiences of Indigenous Women in the Australian Mining Industry

Joni Parmenter

Introduction

Around the world, research has shown that the introduction of large-scale mining adversely affects women in indigenous communities to a greater extent than men (Tauli-Corpuz 1997; Bhanumathi 2003; Bose 2004). A major factor that has contributed to adverse impacts experienced by women is that they have largely been excluded from negotiations concerning benefits from mineral development, including employment (Connell and Howitt 1991; Gibson and Kemp 2008). Organisations such as the World Bank and Oxfam now recognise the potential disadvantage experienced by women. They now insist on the inclusion of gender aspects in impact assessments and promote gender equality as 'smart economics' (World Bank 2006). There is a substantial body of empirical evidence to demonstrate that the social and economic empowerment of women contributes to economic growth, poverty reduction, effective governance and more sustainable development in local communities (World Bank 2001: 1).

Not all impacts of mining on women are considered adverse. There is some evidence of positive outcomes for women, such as increased access to education and travel (Robinson 2002: 43), and benefits accrued via improvements in infrastructure, such as roads and transport, which can provide access to new markets (Byford 2002). Another way women might experience benefits from mining is through employment. For a long time, mining has been considered a very masculine industry by virtue of its heavily male-dominated workforce and the physicality of mining work. Women who do gain employment in the mine are often treated with 'condescending chivalry' or treated as a novelty (Miller 2004: 49). Further, the sexist views faced by women entering the mining workforce often limit career advancement (Tallichet 2000; Gibson and Scoble 2004).

Despite a long history of working in the mining sector, sometimes performing harder or more work then their male counterparts (Amutabi and Lutta Mukebi 2001), women miners have been 'hidden from history' (Burke 2006). There is a growing body of literature on the experiences of indigenous women working in small-scale and artisanal mining in developing countries (Lahiri-Dutt and Macintyre 2006), but very few sources detail their experiences in large-scale mines. Similarly, to date, there has been very limited published information on the experiences of indigenous women in the Australian mining industry. Although there have now been a few studies about female employment in the mining industry in Australia (Pattenden 1998; Kemp and Pattenden 2007) these tend to focus on the experience of non-indigenous women, who make up the majority of female employees.

Increasing the participation of women is high on the agenda of the Australian minerals industry, driven in large part by widespread skill shortages and the need to maximise the human resource pool. Concurrent with this is another, separate agenda to increase the overall participation of indigenous people in the industry (Parmenter and Kemp 2007). Because indigenous women represent the overlapping intersection of these two agendas, the specific needs of indigenous women are at risk of not being recognised or understood.

This chapter brings visibility to indigenous women in the Australian mining industry, by drawing on research conducted with indigenous female employees at Century Mine in northwest Queensland. Where relevant, reference is also made to data collected by government agencies such as the Australian Bureau of Statistics (ABS) and Australian Bureau of Resource Economics (ABARE). This chapter does not attempt to provide a comprehensive analysis of the experience of indigenous female employees in mining in Australia as there is still only limited data available. The chapter is divided in three sections. The first section presents an overview of participation in the Australian mining industry, the second section provides a case study of indigenous women working at Century Mine and the third section discusses the theoretical background and implications for incorporating indigenous women in gender mainstreaming practice. The chapter concludes by arguing that gender should not be considered in isolation to other intersecting factors such as race, and that understanding the experience of indigenous women in the mining workplace is important in contributing to long-term positive outcomes for indigenous communities.

Participation in the Australian Mining Industry

Indigenous women have worked in the mining industry in Australia since at least the 1940s (Wilson 1961, 1980, cited in Holcombe 2004). There has been a steady increase in indigenous female participation in the Australian minerals industry over time, but more sharply in recent years. One explanation for this is that Australia has experienced a resources boom, and another is the recent efforts of some mining companies to diversify the workforce. The data presented ahead provides some general insight into the participation of indigenous women currently working in the Australian minerals industry.

The Australian 2006 National Census identified 395 indigenous women working in mining, representing 0.4 per cent of the total mining workforce. This figure represented 16 per cent of the indigenous mining workforce, compared with a 15 per cent representation of non-indigenous women in the non-indigenous mining workforce. There has been an increase in indigenous women's representation when compared to the 2001 Census, which identified 156 indigenous females working in mining, representing 0.2 per cent of the total workforce and 11.2 per cent of the minerals industry's indigenous workforce. When viewed by commodity, the coal sector has by far the lowest overall indigenous employment rate, at only 0.3 per cent of the sector workforce (Tedesco et al. 2003). This can be explained in part by the fact that coal mining regions have a lower representation of indigenous people in the population than metalliferous mining regions. The coal sector also has the lowest representation of all women, at only 4 per cent of the sector workforce (Kemp and Pattenden 2007). In interpreting these figures it is important to note that they are likely to under-represent the true number of indigenous people due to the reluctance of some indigenous people to self-identify, and the ongoing issue of variable levels of indigenous participation in the National Census (Taylor and Bell 2004).

A recent study of indigenous employment in the Australian minerals industry, conducted by Tiplady and Barclay (2006), identified 243 indigenous women representing 2.2 per cent of the total workforce across 12 participating sites.[1] Women accounted for 20 per cent of the total indigenous workforce, although there was considerable variation across sites, with female representation ranging between 8 to 33 per cent of the total indigenous workforce. The majority of male and female indigenous employees in the Indigenous Employment Study worked in semi-skilled positions (57 per cent) (see Table 5-1).

---

1   These sites were selected primarily on the basis that the operation and/or parent company had shown a commitment to increasing indigenous employment and would, therefore, be more likely to have higher participation rates than other sites in Australia. The largest number of case study sites were in Western Australia (seven) followed by Queensland (three) and the Northern Territory (two) (Tiplady and Barclay 2006).

Table 5-1: Occupations of indigenous employees in the Indigenous Employment Study.

Occupation Type	Men	Women	Total	Indigenous workforce (%)
Semi-skilled	578	89	667	56.6
Administration	4	73	77	6.5
Professional	7	7	14	1.2
Trade	82	3	85	7.2
Supervisor	31	2	33	2.8
Technical	19	2	21	1.8
Superintendent	8	0	8	0.7
Specialist	5	2	7	0.6
Manager	5	0	5	0.4
Graduate	1	1	2	0.2
Executive/Manager	0	0	0	0
Traineeship	100	60	160	13.6
Apprentice	96	4	100	8.5
Total	**936**	**243**	**1179**	

Source: Adapted from Tiplady and Barclay (2006).

A similar situation is seen throughout history and in other areas of the world. In India, the *adivasi* (indigenous people) of the collieries were concentrated in low positions (Sinha 2002), and in South Africa, professional women in mining are more likely to be white, while unskilled women in mining are more likely to be black (Ranchod 2001: 6). Excluding the indigenous women in semi-skilled positions (for example, truck drivers), the categories of jobs undertaken by indigenous women tend to be aligned with those usually associated with women.[2] There were significantly fewer indigenous female apprentices, tradespersons, supervisors and technical staff, and significantly more women in administration roles. A similar pattern is seen in the female mining workforce more broadly (Kemp and Pattenden 2007).

The lack of participation of indigenous women in the Australian mining industry has been attributed to several factors, some of which are not unique to indigenous women. Indigenous women are more likely to bear children at a younger age than non-indigenous women, and are often responsible for caring for larger numbers of dependents or others than non-indigenous women (ABS 2008). A young family may be a constraint in terms of starting a career; however, indigenous women often have a close extended family that may

---

2 'Female' roles are those work roles usually associated with women, including administration, catering, cleaning and professional support roles such as Human Resources, public relations, and community relations. Roles usually associated with men, are mining engineers, metallurgists and operational roles, such as truck driving (Kemp and Pattenden 2007).

care for the children while they work. Dominant values in some indigenous communities regarding women's involvement in waged employment may also influence an individuals' desire to participate (O'Faircheallaigh 1998). Other cultural factors attributed to lack of participation of indigenous Australians include the availability of hunting and gathering activities in preference to jobs and whether a person's first language is English or indigenous (Hunter and Gray 1999).

The systemic societal disadvantage experienced by indigenous communities raises many complex issues. Lack of education and poor health directly affects employment prospects for indigenous people. Indigenous Australians are more likely than non-indigenous Australians to suffer from asthma, diabetes and cardiovascular disease, and are also more likely to report health risk factors such as smoking and excessive drinking (Tiplady and Barclay 2006). However, according to census data, indigenous women are less likely than indigenous men to engage in drug and alcohol abuse (a significant safety issue within the mining industry) and are also likely to be more educated than indigenous men (ABS 2006).

Two or three individual mines in Australia have been able to achieve indigenous workforce representation of up to 20 per cent of the total workforce. One of these mines is Century Mine, where the data for this chapter was collected. The next section provides an overview of Century Mine, and presents results from recent interviews with indigenous women working there.

Century Mine and Indigenous Female Employees

Mineral Metals Group's (MMG) Century Mine is a large open cut zinc mine located in the lower Gulf of Carpentaria region of North West Queensland (see Figure 5-1).

The relationship between Century Mine and adjacent communities in the Gulf of Carpentaria is generally mediated by the Gulf Communities Agreement (GCA). This is a right-to-negotiate (RTN) agreement made between Pasminco Century Mine Limited,[3] (now MMG), the Queensland Government and representatives from the Waanyi, Mingginda, Gkuthaarn and Kukatj native title groups, signed under the provisions of the Commonwealth Native Title Act 1993 on 13 February, 1997. Increasing indigenous employment is a cornerstone of this agreement, however, employment is not defined in terms of gender considerations.

---

3   In September 1997 Pasminco purchased the Century Mine project from Rio Tinto and in March 2004 Pasminco was relaunched on the stock exchange as Zinifex. In 2008 Zinifex merged with Oxiana and re-launched on the stock exchange as OZ Minerals. In 2009, OZ minerals were sold to Minerals and Metals Group (MMG).

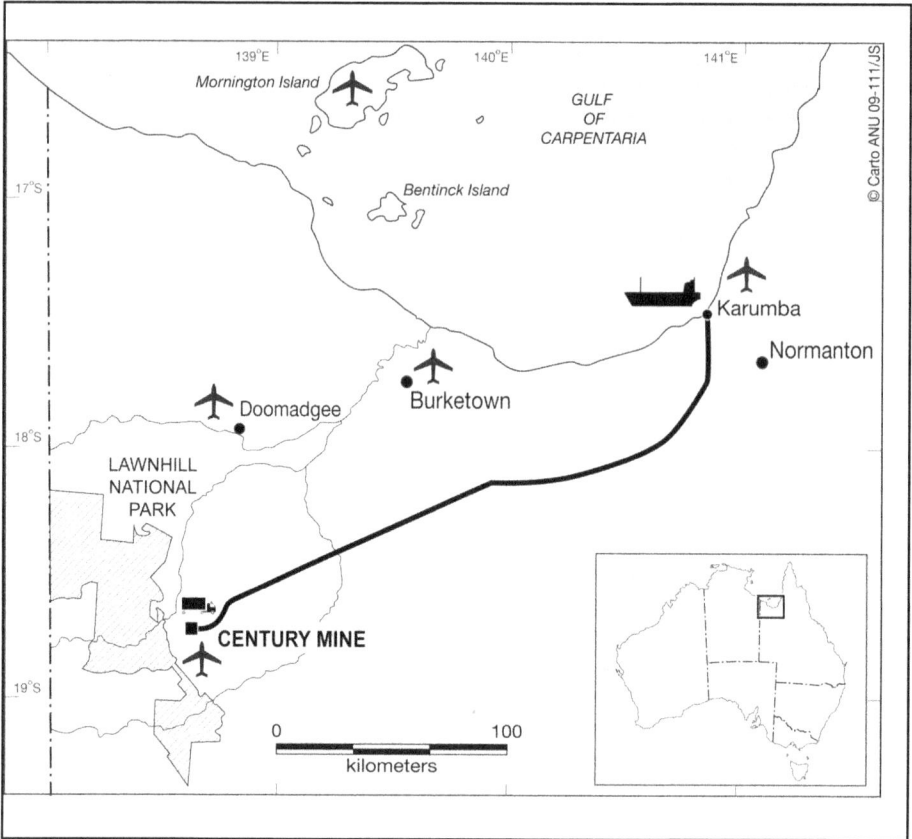

Figure 5-1: Map of Century Mine and surrounding gulf communities.

Source: Century Mine.

At the time of writing, Century Mine employed approximately 1130 employees, of whom 207 (18 per cent) are indigenous. The participation rate for both indigenous people and indigenous women is relatively higher than most other mine sites in Australia. This can be attributed partly to the fact that the mine is located in a region where there are high populations of indigenous people, and because the Gulf community's agreement includes provisions to employ indigenous people locally.

According to data supplied by Century Mine, in September 2008 there were 48 indigenous female employees working at Century Mine, representing 4.2 per cent of the total workforce and 23 per cent of the total indigenous workforce (see Table 5-2). As with indigenous men, the majority of these positions are categorised as semi-skilled; mostly as truck drivers. Jobs are aligned with those typically associated with women, with all indigenous people in administration positions occupied by women and all trades and apprenticeships occupied by indigenous men. There were few indigenous men and no indigenous women in supervisory or management roles.

Table 5-2: Indigenous employee positions at Century Mine: September 2008.

Occupation Type	Male	Female	Total
Semi Skilled	109	28	137
Trade	16	0	16
Administration	0	5	5
Manager	1	0	5
Supervisor	4	0	4
Technical	1	2	3
Superintendent	1	0	1
Specialist	0	0	0
Professional	0	0	0
Apprentice	12	0	12
Traineeship	7	8	15
Unidentified	8	5	9
Total	**159**	**48**	**207**

Source: Author's data.

Indigenous female employees at Century Mine live in a variety of communities and towns in northern Queensland. Some live more locally in the Gulf communities and others fly in from the larger towns of Mt Isa and Townsville. Based on interviews undertaken with 36 indigenous female employees at Century Mine, women from the Gulf communities, particularly Doomadgee and Mornington, find coming to work at Century more challenging than indigenous women from the other towns, who have more exposure to mainstream employment and lifestyles. The indigenous population represents over 90 per cent of the total population in Doomadgee and Mornington Island, both of which are disadvantaged socially and economically. For many of these women, it is not only their first experience in mainstream employment but also their first experience in a non-indigenous, male dominated environment. Indigenous women from the local Gulf communities make up 44 per cent of the total indigenous female workforce. The remaining 56 per cent live in the larger towns of Mount Isa and Townsville.

The following discussion of indigenous women's experience working at Century Mine draws on field work conducted in September 2008. Nine women were interviewed on a one-on-one basis; and three focus groups and two group interviews were undertaken. Interviews were semi-structured. Thirty-six indigenous women participated in total, representing 75 per cent of the indigenous women currently working at Century. Participants' ages ranged from 19 to 51 years and their periods of employment ranged from five weeks to nine years. The women interviewed worked in varied roles and areas across the site, including cleaners and kitchen hands in the village (accommodation area) and operators in the open cut pit. Some pre-vocational trainees were also interviewed. This discussion also draws on previous studies conducted by The University of Queensland's Centre for Social Responsibility in Mining (CSRM) in 2006 and 2007 involving indigenous women at Century Mine. The previous fieldwork was conducted during three visits to the mine site and surrounding communities at different times between 2005 and 2007. One study focused on community perspectives, one on the implications of closure for Gulf communities, and one on the experiences of women in mining. Multiple methods were used, including a quantitative survey, semi-structured interviews and focus groups. Some of the same indigenous female employees participated in more than one study.

Indigenous Women at Century Mine: The Experience

Cultural Constraints

Working in the mining industry may not be considered culturally appropriate for some indigenous people (Trigger 2002), whereby a responsibility exists to care for 'country'. The word 'country' was adopted by English-speaking aboriginal peoples to describe their reciprocal relationship to homelands. If you are doing the right thing ecologically, the results will be social and spiritual, and vice versa (Rose 2002: 49). Some indigenous people may therefore refrain from working in mining for this valid cultural reason. However, not surprisingly, given their employment at the mine, none of the women interviewed at Century Mine felt compromised by this factor.

For some indigenous employees, 'avoidance rules' that dictate whether individuals may communicate with each other may be problematic in the workforce environment. For example, two people who are usually not permitted to speak to each other may be required to work in the same crew, be on the same flight or supervise one another. According to indigenous women at Century, this situation does occur occasionally and management has worked around it. However, this is an informal process, so some incidents may go undetected. It is unknown if these 'avoidance rules' have a greater effect on women or men. In regard to supervision, one indigenous woman explained that age was more of an issue than gender, with older indigenous people of both genders less likely to take orders from younger people of both genders. This needs further research.

In addition, a gender-based cultural constraint creates a very practical challenge for indigenous women at Century Mine. Among some indigenous groups in northern Australia, red ochre (hematite) is specifically the domain of men (it is important in male ceremonial contexts) and it is considered offensive and dangerous for women to touch, or be in close contact with. This clearly makes it very problematic for women to work in the open cut pit, where this material is uncovered regularly. At the request of traditional owners, Century Mine management does not allow women to work directly with ochre. Women (including non-indigenous women) are not permitted to haul the material in their trucks, and it has to be 'sheeted'[4] before a woman may drive over it. Although the majority of indigenous people on site appeared to respect this cultural tradition, some non-indigenous women operators have expressed frustration. The indigenous women on the other hand, strongly support the cultural significance of this material and the associated exclusion rules. At the time of the interviews, there was a large amount of ochre in the open pit, in an area known as the 'west'. As a result, women were assigned tasks such as painting tyres, or moved over to work in an area known as the 'kingdom'. The 'kingdom' was so nicknamed due to a popular previous supervisor of ore mining, who identified the area at the time as his own; it was his 'kingdom'. When the area moves, so to does the name, which now generally refers to the area whose ore is being mined.

According to the indigenous women who work as truck drivers or water cart operators, working in the kingdom is harder work. It is a slower run, and very difficult for fatigue management. Although there is little evidence to suggest that men are abusing this cultural tradition, either to make women perform harder work or to exclude them from the pit altogether, this remains an area for further investigation. For example, at one mine site in Papua New Guinea (PNG), men have gone to extraordinary lengths to exclude women from operating the dump trucks. Objections were raised in terms of the danger of pollution: to men who had to sit in the same seat where a menstruating women had sat; to the vehicle that would be infused with dangerous emanations from female bodies and to the motor mechanics who would be exposed to the polluted sump oil that would ultimately be the repository for these emanations (Macintyre 2003: 5–6). Despite opposition from women, the mine decided not to allow women on dump trucks as it would be 'against custom' (Macintyre 2003).

There may also be implications of working at the mine that affect indigenous women in their home environments. For example, it is unknown if humbugging[5] for mining income affects women more or differently than men. At home, women from the communities of Doomadgee and Mornington Island said they

---

4   Sheeting refers to the laying down of material on the pit floor, usually limestone or shale.
5   'Humbugging' refers to pressuring a relative for money.

felt added pressure to share their money (humbug) since working at the mine. One woman said: 'Our money is not our money, we gotta share it … aboriginal way. Pressure to share with kin or 'demand sharing' is noted in the literature for both genders (Petersen 1993; Trigger 2005) and is important in the constitution of social relations in egalitarian societies. Currently little is known about how income from mining employment is distributed by indigenous workers and how many people benefit. Further, as O'Faircheallaigh (1998) pointed out over a decade ago, there is a need to determine whether waged employment will sharpen or reduce inequality between indigenous men and women.

Extra Burdens

Some of the experiences of the indigenous women interviewed are not unique to indigenous women, such as the pressures of family responsibilities and issues associated with working in such a male-dominated environment (Kemp and Pattenden 2007). However, these issues can be compounded for indigenous women, who often have additional cultural or extended family responsibilities, and are more likely to come from socio-economically disadvantaged communities than non-indigenous women. Further, indigenous women may endure the extra burden of experiencing racism as well as sexism. This may explain why the great majority of indigenous women at Century Mine perceived themselves as occupying the bottom position of the mine site 'hierarchy', where non-indigenous men are at the top, followed by non-indigenous women, indigenous men and then them. The women are both indigenous and female—a double minority resulting in the bottom position.

Many indigenous women at Century spoke of the extra burden of having family responsibilities and the need to 'prove yourself' at a mine site. One woman said:

> The most annoying thing is that women work hard and they are still the mothers, they still got to do all the work at home and come up here and have to do more work than men because they have to prove themselves, its even harder for an indigenous woman, feel like you're being monitored more.

The indigenous women who worked in the open cut pit felt more pressure from their male colleagues compared to women in other areas on site, where women were represented in greater numbers. One woman from the pit said:

> Can't be weak, can't take everything to heart, have to be able to relate to men … they shit-stir you, try to make you break, try to make you leave … that's their idea of having a good time.

If women showed any form of weakness they were reminded of their 'femaleness'. For example, a group of women agreed that whenever there was a circumstance when they had to leave their shift due to illness, responses from men included: 'You're going home early because you are a woman, if you were a man you'd stay', whereas nobody said anything to male workers in the same situation.

Many of the women thought it was harder for women to get promoted than men, and even harder for indigenous women. One woman said: 'Mining has always been a men's thing. We are the cleaners, we never get anywhere else. Men can always get in before a woman because it's mining'. A few women who had some experience in a temporary supervisory role, said that the men they supervised were unwilling to take orders. One woman said: 'Men don't like taking orders from a woman, if a bloke[6] tells them they go out and do it straight away, but if we do they don't'. Another woman told a story of how she managed this 'push back' from men:

> I wrote instructions for the day on the board, rather than tell them directly, if I did have to speak to them, I'd start with: 'Would you mind doing me a favour?'

In a previous CSRM study that involved indigenous women at Century Mine (Kemp and Pattenden 2007), one woman commented on how she thought non-indigenous supervisors perceived indigenous female employees:

> They don't take you seriously ... they're not going to listen to a woman, think a woman can't know any better than they do, especially a black woman.

Indigenous women in the current study confirmed this position. One woman said:

> White women get looked after—do they think we are not intelligent enough to move up to these positions? I mean we're not back in the Stone Age, there are some smart aboriginals out there.

Almost all interviewees said they felt they were overlooked for promotions, or less favoured by supervisors (who are almost entirely non-indigenous males) because they were both indigenous and female, and therefore 'especially dumb'. One woman noted that in reality, indigenous women were more likely to be more highly educated than indigenous men in their community. This is confirmed by ABS statistics (ABS 2006).

All indigenous women interviewed felt that non-indigenous women were favoured for career advancement and generally more supported by non-

---

6   'Bloke' is a colloquial expression for 'man'.

indigenous male supervisors. One woman commented that unlike new indigenous female employees, new non-indigenous female employees in the open cut pit were made to feel welcome. Their names were known within a few days, with supervisors congratulating and flirting with them on the mines communication channel: 'It's like ... "welcome to our group," sort of thing.' The indigenous women also thought that non-indigenous women were more aggressive, more likely to complain and, therefore, more likely to get promoted or secure training opportunities than indigenous women. They said indigenous women were less likely to complain and 'just do what they're told' and were therefore more likely to be taken advantage of and allocated the unpleasant jobs.

The majority of women at Century self-identify with being a member of an oppressed racial group, before being female. This is not to say that indigenous men are innocent of the same chauvinist attitudes that non-indigenous women have accused non-indigenous men of holding (Huggins 1987). A few of the women interviewed said that working at the mine has enabled them to have more independence from their husbands, who they perceived to be too controlling. One woman said:

> Before I worked, my husband expected me to stay home with the kids all day while he was at work, and then all night as well while he went to the pub ... now we share.

Indirectly, indigenous men were criticised by many of the women for what was termed 'jealousy'. Many women said that this jealousy was the main barrier preventing indigenous women from working at the mine. People in indigenous communities did not want their partners to come and work for the mine, assuming there are too many opportunities for infidelity. Although this jealousy is apparently felt by both genders, clearly it would affect more women given that the mine site is so male dominated. One woman said that this jealousy is such a major factor, that women in her community had previously been forced to choose between their partner and their job. This jealousy also prevented women from attending the wet mess (bar) on site, in order to avoid married men and starting rumours back in the community. Some also thought it was uncomfortable visiting the bar because of the perceptions held by non-indigenous men: '[T]hey just stare at you, assuming you are on the piss ... you feel preyed upon'.

The majority of indigenous women interviewed felt strongly about being 'looked down on' by both non-indigenous men and women, but were much less critical of indigenous men, seemingly preferring to 'stick together' as indigenous people, against what they saw as the more dominant oppressor—white men and women. As Huggins (1987) stated 20 years ago, it is easier for an indigenous woman to consider herself first a human being, second an indigenous person and third a member of the female sex, viewing disadvantages of race and class before those of sex. This relates to the indigenous feminist literature that recognises the intersection of gender, race and colonialism (Green 2007), discussed ahead.

Gender Mainstreaming and Intersectionality

There is now general agreement that the mining industry is highly gendered and the negative impacts of mining are greater for women than for men (Tauli-Corpuz 1997; Bhanumathi 2003; Bose 2004). Less is known about the relationship between other intersecting factors such as race and class between men and women and between women as a group. The relationship between gender inequality and other complex inequalities is an important and unresolved debate in both feminist theory and gender mainstreaming[7] practice (Walby 2005: 463). The Century case indicates that some female aboriginal mine workers view issues of race as more important than those of sex. This is further supported by African American feminists of the 1970s (Spelman 1988) and the emerging literature on indigenous feminism (Green 2007).

Feminism is a complex subject that has not only redefined itself throughout history but continues to be debated. The commonality between all the different forms of feminism and there are several—is that they have all arisen out of conditions of patriarchy. Therefore, for women who do not perceive their society to be patriarchal, or do not believe they experience oppression from a patriarchal society, feminism is considered irrelevant (Green 2007). This is the case for some indigenous women, who argue that male domination is not universal (Turpel 1993; Monture-Angus 1995). Further, as Monture-Angus (1995) attests, accepting the western idea of 'equality' would mean accepting a lower position than what has historically been accorded to women in her culture. Some Australian indigenous women have expressed similar views towards the women's movement (Huggins 1994; Moreton-Robinson 2000). In 1994, Huggins suggested there was no support for the women's movement from within indigenous circles. Ten years later, Fredericks (2004) wrote: 'We stand and watch non-indigenous women argue for something which we had and which they assisted in disempowering us of'. When western women were addressing their sexual oppression by men *via* the feminist movement, indigenous women

---

7   Gender mainstreaming is the process of assessing the implications for women and men of any planned action, including legislation, policies and programs, in all areas and at all levels. It is a strategy for making women's as well as men's concerns and experiences an integral dimension of the design, implementation, monitoring and evaluation of policies and programs in all political, economic and societal spheres so that women and men benefit equally and inequality is not perpetuated. The ultimate goal is gender equality (UN ECOSOC 1997: 2).

(and men), were addressing many forms of oppression by the dominant western society. Therefore, indigenous women are coming from a very different position than western women.

Green (2007) argues that whether or not indigenous women have experienced patriarchal oppression within their own cultures and communities, they are subjected to the patriarchal and colonial oppression within the dominant western society. There is now a small body of literature emerging on indigenous feminism, as evidenced by Green's 2008 work, titled *Making Space for Indigenous Feminism*. Some indigenous scholars who once rejected feminism are now acknowledging its importance and identifying themselves as indigenous feminists (Bear 2007; St Denis 2007). Indigenous feminists raise issues of colonialism, racism and sexism, and the unpleasant synergy between these violations of human and political rights (Green 2007: 20). In the case of mining, this synergy is likely to be further exacerbated for indigenous female employees given that the Australian mining industry is western, masculine, and heavily dominated by non-indigenous male workers.

If race and class have historically come before gender for indigenous women as barriers to their advancement, the question remains as to how and if gender mainstreaming is appropriate for indigenous women, and if so, how it can be implemented in the mining industry. Gender mainstreaming has been widely adopted by international aid agencies for some time as a key strategy for achieving gender equality. However, for the mining industry, it is a relatively new concept. With the advent of the United Nations (UN) Millennium Development Goals (MDG), the mining industry has taken note. The MDG assigns collective responsibility for halving world poverty by 2015, with an explicit commitment to gender equality as an end in itself. In 2009, the mining company Rio Tinto produced a resource guide for integrating gender considerations into community work at their operations. This is one of the first publicly-articulated, community-oriented gender frameworks developed by a multinational mining company. However, indigenous women and the intersection between race and gender is not a focus.

Gibson and Kemp (2008) use the term 'double blind' to describe corporate engagement that tends to overlook the nexus between gender and indigeneity. There is little evidence that the industry's agenda to increase the employment of indigenous people includes a gender dimension, and little evidence that the agenda to increase the participation of women includes an indigenous component. At a site level, there is some evidence of consideration for indigenous cultural issues, such as allowing leave for funerals and compulsory cultural awareness training for all employees, but these are not gender specific. In this sense, the

Australian mining industry has, not unlike western feminist theory and gender mainstreaming practice, failed to consider the intersection between race and gender.

Current debate exists about whether it is preferable to adopt the concept of diversity mainstreaming. Both gender mainstreaming and diversity mainstreaming are contested concepts. Some have argued that the concept of gender is invariably tied to the male-female binary and therefore limited in its ability to reflect differences among women (Hankivsky 2005, Squires 2005, Walby 2005). Take for instance indigenous women's ready acceptance of gender exclusion rules that apply with the management of red ochre in the Century pit. Unlike non-indigenous women, there is acceptance because the exclusion is culturally based and shows identification with their aboriginality taking precedence over any concerns about gender inequality. Others are concerned that by putting all inequalities together under the one umbrella of 'diversity', risks overlooking political intersectionality and may cause dilution of the feminist agenda (Mackay and Bilton 2003; Verloo 2006).

In a recent gender analysis study that produced guidelines for the Western and South Australian state public sectors, indigenous female representatives expressed strong reservations about the usefulness of the concept 'gender' (Bacchi and Eveline 2009). Gender was understood to privilege male/female relations and was considered problematic for their social analysis of racialisation (ibid.). Further, using the very word 'mainstreaming' proved difficult due to its association with political moves to disband the democratically elected representative body of Aboriginal and Torres Strait Islander Commission (ATSIC) (Pratt and Bennett 2004–2005). The issue was resolved by introducing 'race and cultural analysis' as informing gender analysis. Race and cultural analysis requires the whole discussion of equality and equity to be rethought through the perspective of aboriginal people (Bacchi and Eveline 2009). Although Bacchi and Eveline (ibid.) do not suggest introducing this analysis as a model for developments elsewhere, the need to respond to on-the-ground political developments has merit. Gender mainstreaming for the mining industry may need to consider a non-generic approach to implementation that also reflects the current level of theoretical development concerning the question of gender and its intersecting relations.

Conclusion

This chapter has discussed key findings from recent research that substantially involved indigenous women working at Century Mine. The research has highlighted how the challenges faced by women working in the Australian mining industry can be compounded for indigenous women due to additional familial and cultural responsibilities. For many indigenous women who live in remote communities, this is their first experience in mainstream employment, in an environment that is heavily dominated by non-indigenous and mostly male workers. Indigenous women are both indigenous and female, a double minority in a very unfamiliar environment, thereby conflating the experiences of both sexism and racism.

On a global level, it is important that the resources industry does not consider gender alone, but includes other intersecting identities and factors such as race, class and the local socio-political and cultural context of the women affected. The Australian mining industry has so far failed to consider the intersection between race and gender. The industry's agenda to increase the participation of women in the workforce does not include an indigenous component, and the agenda to increase indigenous participation in the workforce does not include a gender component.

This chapter is a first step in bringing visibility to the experience of indigenous female employees in the Australian mining industry. Although this research provides insight into some indigenous women's experiences in mining employment, further dedicated research focusing on indigenous women in mining is warranted. If mining companies are to attract and retain more indigenous women, as well as contribute to the long term sustainability of indigenous communities, it is crucial that they understand and respond to the impacts of employment more fully.

References

Amutabi, M. and M. Lutta Mukebi, 2001. 'Gender and Mining in Kenya: The Case of Mukibirs Mines in Vihiga District.' *Jenda: A Journal of Culture and African Womens Studies* 1(2): 1–23.

ABS (Australian Bureau of Statistics), 2006. 'Census of Population and Housing.' Canberra: Australian Bureau of Statistics.

———, 2008. 'Australian Social Trends.' Canberra: Australian Bureau of Statistics.

Bacchi, C. and J. Eveline, 2009. 'Gender Mainstreaming or Diversity Mainstreaming? The Politics of "Doing".' *NORA: Nordic Journal of Feminist and Gender Research* 17(1): 2–17.

Bear, S., 2007. 'Culturing Politics and Politicising Culture.' In J. Green (ed), op.cit.

Bhanumathi, K., 2003. 'Women and Mining in India.' Paper presented at conference on 'Women in Mining', Melbourne, 3 August.

Bose, S., 2004. 'Positioning Women within the Environmental Justice Framework: A Case for the Mining Sector.' *Gender, Technology and Development* 8(3): 407–12.

Burke, G., 2006. 'Women Miners: Here, Now and Then.' In K. Lahiri-Dutt and M. Macintyre (eds), op.cit.

Byford, J., 2002. 'One Day Rich: Community Perceptions of the Impact of the Placer Dome Gold Mine, Misima Island, Papua New Guinea.' In I. Macdonald and C. Rowland (eds), op.cit.

Connell, J. and R. Howitt (eds), 1991. *Mining and Indigenous Peoples in Australasia*. Sydney: Sydney University Press.

Fredericks, B., 2004. 'A Reconstituted Social Environment: Feminism and the Plight of Aboriginal Women in Australia.' *Newtopia Magazine, Ideological Issue* 3(18).

Gibson, G. and D. Kemp, 2008. 'Corporate Engagement with Indigenous Women in the Minerals Industry: Making Space for Theory.' In C. O'Faircheallaigh and S. Ali (eds), *Earth Matters: Indigenous Peoples, the Extractive Industries and Corporate Social Responsibility*. Sheffield. Greenleaf Publishing.

Gibson, G. and M. Scoble, 2004. 'Regendering Mining: A Survey of Women's Career Experiences in Mining' *CIM Bulletin*, 97(1082): 55–61.

Green, J., 2007. 'Taking Account of Aboriginal Feminism.' In J. Green (ed.), op.cit.

Green, J., (ed.), 2007. *Making Space for Indigenous Feminism*. Canada: Fernwood Publishing

Hankivsky, O., 2005. 'Gender vs Diversity Mainstreaming: A Preliminary Examination of the Role and Transformative Potential of Feminist Theory.' *Canadian Journal of Political Science* 38(4): 977–1001.

Holcombe, S., 2004. 'Early Indigenous Engagement with Mining in the Pilbara: Lessons from a Historical Perspective.' Canberra: Centre for Aboriginal Economic Policy Research (Working Paper 24). Viewed 7 March 2006 at http://www.anu.edu.au/caepr/Publications/WP/CAEPRWP24.pdf

Hunter, B. and M. Gray, 1999. 'Further Investigations into Indigenous Labour Supply: What Discourages Discouraged Workers?' Canberra: Centre for Aboriginal Economic Policy Research (Working Paper 2).

Huggins, J., 1987. 'Black Women and Women's Liberation.' *Hecate: An Interdisciplinary Journal of Women's Liberation* 13(1): 77–82.

———, 1994. 'A Contemporary View of Aboriginal Women's Relationship to the White Women's Movement.' In N. Grieve and A. Burns (eds), *Australian Women and Contemporary Feminist Thought*. South Melbourne: Oxford University Press.

Kemp, D. and C. Pattenden, 2007. 'Retention of Women in the Minerals Industry.' In Centre for Social Responsibility in Mining Team (eds), *Unearthing New Resources: Attracting and Retaining Women in the Australian Minerals Industry*. Canberra: Australian Government Office for Women, Minerals Council of Australia, Women in Social and Economic Research and Centre for Social Responsibility in Mining. Viewed 6 September 2010 at http://www.csrm.uq.edu.au/docs/women-in-mining.pdf

Lahiri-Dutt, K. and M. Macintyre (eds), 2006. *Women Miners in Developing Countries: Pit Women and Others*. Aldershot: Ashgate.

Macdonald, I. and C. Rowland (eds), 2002. *Tunnel Vision: Women, Mining and Communities*. Melbourne: Oxfam Community Aid Abroad.

Macintyre, M., 2003. 'The Changing Value of Women's Work on Lihir.' Paper presented at conference on 'Women in Mining: Voices for Change.' Madang, 3–6 August.

MacKay, F. and K. Bilton, 2003. 'Learning from Experience: Lessons in Mainstreaming Equal Opportunities.' Edinburgh: University of Edinburgh, Governance of Scotland Forum. Viewed 6 September 2010 at http://www.scotland.gov.uk/Publications/2003/05/17032/21487

Miller, G., 2004. 'Frontier Masculinity in the Oil Industry: The Experience of Female Engineers.' *Gender, Work and Organization* 11(1): 47–73.

Monture-Angus, P., 1995. *Thunder in My Soul: A Mohawk Woman Speaks*. Halifax: Fernwood Publishing.

Moreton-Robinson, A., 2000. *Talkin' Up to the White Women: Indigenous Women and Feminism*. St. Lucia: University of Queensland Press.

O'Faircheallaigh, C., 1998. 'Resource Development and Inequality in Indigenous Societies.' *World Development* 26(3): 381–94.

Parmenter, J. and D. Kemp, 2007. 'Indigenous Women in Mining Employment in Australia.' Paper presented at conference on 'Sustainable Development.' Cairns, 29 October–2 November.

Pattenden, C., 1998. 'Women in Mining: A Report to the "Women in Mining" Taskforce of the AusIMM.' Melbourne: AusIMM.

Peterson, N., 1993. 'Demand Sharing: Reciprocity and the Pressure for Generosity among Foragers.' *American Anthropologist* 95(4): 860–74.

Pratt, A. and S. Bennett, 2004–2005. 'The End of ATSIC and the Future Administration of Indigenous Affairs.' Canberra: Parliamentary Library (Current Issues Brief 4, 2004–2005). Viewed 17 July 2010 at http://www.aph.gov.au/library/pubs/CIB/2004-05/05cib04.htm

Ranchod, S., 2001. 'Gender and Mining: Workplace.' Johannesburg: African Institute of Corporate Citizenship. Viewed 10 June 2008 at http://www.naturalresources.org/minerals/cd/mmsd_saf.htm

Robinson, K., 2002. 'Labour, Love and Loss: Mining and the Displacement of Women's Labour.' In I. Macdonald and C. Rowland (eds), op. cit.

Rose, D.B., 2002. *Country of the Heart: An Indigenous Homeland*. Canberra: Aboriginal Studies Press.

St Denis, V., 2007. 'Feminism is for Everybody: Aboriginal Women, Feminism and Diversity.' In J. Green (ed.), op. cit.

Sinha, S., 2002. 'Colonialism and Capitalism: Unearthing the History of Adivasi Women Miners of Chotangapur.' In K. Lahiri-Dutt and M. Macintyre (eds), op. cit.

Squires, J., 2005. 'Is Mainstreaming Transformative? Theorising Mainstreaming in the Context of Diversity and Deliberation Social Politics.' *International Studies in Gender, State and Society* 12(3): 366–88.

Spelman, E., 1998. *Inessential Woman: Problems of Exclusion in Feminist Thought*. Boston: Beacon Press.

Tallichet, S.E., 2000. 'Barriers to Women's Advancement in Underground Coal Mining.' *Rural Sociology* 65(2): 234–53.

Tauli-Corpuz, V., 1997. 'The Globalization of Mining and its Impact and Challenges for Women', Paper presented at conference on 'Women and Mining', Baguio City, 20–8 January.

Taylor J. and M. Bell, (eds), 2004. *Population Mobility and Indigenous Peoples in Australasia and North America*. London: Routledge.

Tedesco, L., M. Fainstein and L. Hogan, 2003. 'Indigenous People in Mining.' Canberra: Australian Bureau of Agricultural and Resource Economics (eReport 03.19). Viewed 12 September 2007 at http://abareonlineshop.com/product.asp?prodid=12599

Tiplady, T. and M. Barclay, 2006. 'Indigenous Employment in the Minerals Industry.' Brisbane: Centre for Social Responsibility in Mining. Viewed 17 July 2010 at http://www.csrm.uq.edu.au/docs/CSRM%20Report_FINAL%20TO%20PRINT_singles.pdf

Trigger, D., 2002. 'Large Scale Mining in Aboriginal Australia: Cultural Dispositions and Economic Aspirations in Indigenous Communities.' In Proceedings of Council of Mining and Metallurgical Institutions Congress 2002: International Codes, Technology and Sustainability for the Minerals Industry. Carlton: Australasian Institute of Mining and Metallurgy.

———, 2005. 'Mining Projects in Remote Aboriginal Australia: Sites for the Articulation and Testing of Economic and Cultural Futures.' In D. Austin-Broos and G. Macdonald (eds), *Culture, Economy and Governance in Aboriginal Australia*. Sydney: Sydney University Press.

Turpel, M.E., 1993. 'Patriarchy and Paternalism: The Legacy of the Canadian State for the First Nations Women.' *Canadian Journal of Women and Law* 6(1): 174–92.

UN ECOSOC (United Nations Economic and Social Council), 1997. 'Mainstreaming the Gender Perspective into all Policies and Programs in the United Nations System.' Geneva: United Nations.

Verloo, M., 2006. 'Multiple Inequalities, Intersectionality and the European Union.' *The European Journal of Women's* Studies 13(3): 211–28.

Walby, S., 2005. 'Introduction: Comparative Gender Mainstreaming in a Global Era.' *International Feminist Journal of Politics* 7(4): 453–70.

World Bank, 2001. *Engendering Development—Through Gender Equality in Rights, Resources and Voice*. New York: Oxford University Press.

———, 2006 'Gender Equity as Smart Economics: A World Bank Group Gender Action Plan', Washington: World Bank. Viewed 17 July 2010 at http://siteresources.worldbank.org/INTGENDER/Resources/GAPNov2.pdf

6. Indigenous Women and Mining Agreement Negotiations: Australia and Canada

Ciaran O'Faircheallaigh

Introduction

Agreements negotiated between mining companies and indigenous communities are increasingly important in setting the terms on which mining occurs on indigenous lands. This is particularly so in Australia and Canada, which are the focus of this chapter, but agreements are also becoming more widespread and important in developing countries. Their proponents view negotiated agreements as offering an opportunity to fundamentally change the distribution of costs and benefits from resource development. They argue that they do this by including provisions that limit the negative environmental, cultural and social effects of mining, and by allowing indigenous people to share in the wealth that mining creates (O'Faircheallaigh 2004: 303–4; Public Policy Forum 2005: 4–7).

There is a growing literature on the negotiation, content and outcomes of agreements between indigenous peoples and mining companies (see for example, AIATSIS 2008). This shows that some agreements are certainly generating large economic benefits for indigenous communities affected by mining, and are allowing indigenous landowners to effectively protect cultural sites and to have a substantial say in the environmental management of mining projects. In other cases agreements appear to offer indigenous groups limited benefits either by way of economic gain or enhanced protection of cultural and environmental values (Flanagan 2002; O'Faircheallaigh and Corbett 2005; O'Faircheallaigh 2006a, 2008).

Given the increasing importance of negotiated agreements and the uneven nature of the benefits they bring, it seems important to focus on the role of indigenous women in agreement making. Little has been written specifically about the role of women; the primary unit of analysis in most of the literature is the 'indigenous community' or the 'indigenous group'. To the extent that the role of women is discussed, the overwhelming impression both in the academic literature and in the publications of non-government organisations (NGOs) and community activists is that indigenous women are excluded from negotiations. The assumption that this is the case may help to explain the lack of a specific focus in the literature on the role of women. But is this assumption warranted? If it is not and indigenous women are in fact involved in negotiations, what roles do they play? Are they sufficiently involved to be able to shape agreements? If so, what is the impact of their involvement and are women contributing to positive outcomes where these occur? Are they sharing in the benefits generated by agreements?

These are not easy questions to answer. Many agreements are confidential, and the negotiations that lead to them even more so. This is in part because of formal confidentiality agreements between indigenous groups and mining companies, but also because communities may not want to reveal how they deal with negotiations, lest this weaken their hand in the future. Even where it is possible for researchers to gain access to formal negotiations, many critical decisions are taken away from the negotiating table, and the way decisions are revealed in a negotiation may indicate little about who was and was not influential in making them.

This chapter is both exploratory and, of necessity, based to some extent on my own work as a negotiator for indigenous communities,[1] though I have also sought to draw on information available in the public domain. I begin by briefly summarising the existing writing on indigenous women and mining negotiations which, I argue, is oversimplified at least as it relates to Australia and Canada. This results in part from limited availability of information on negotiations, but also because too narrow a definition of 'agreement negotiations' is often adopted. I want to argue that this term in fact encompasses a great deal more than what happens across a table as company and community 'negotiators' hammer out an agreement, and that once this is recognised the role of indigenous women is both more complex and more substantial than the literature would suggest. In addition, the role of women very much depends on the specific structures used to prepare for, oversee and undertake the negotiation of agreements, and to conduct the ongoing negotiation that inevitably continues after an agreement is signed. In making this point I want to stress again the specific context I am discussing, that is, aboriginal societies in Australia and Canada. I recognise that the negotiation structures developed by indigenous groups in these countries may be quite different to those in other political and cultural contexts.

---

1   That work spans nearly 20 years and involves assisting groups to prepare for and undertake negotiations in a variety of legal, political and economic contexts including Cape York, Central Queensland, North Stradbroke Island, the Mount Isa region, the Hunter Valley, the Northern Territory, the Kimberley and the Pilbara in Australia and Nunavut in Canada.

Exclusion of Women from Mining Negotiations

The overwhelming impression gained from a review of the existing published information is that women are largely and often entirely excluded from negotiating mining agreements. For instance in a review of corporate engagement with indigenous women in the minerals industry, Gibson and Kemp (2008: 107) note a number of cases in Australia, Papua New Guinea (PNG), Indonesia and India where women were inadequately represented in or excluded from negotiations. They also note that inclusion on negotiating teams does not mean that women are free to contribute equally. They finish their brief discussion of the topic by noting the limited extent of information on negotiations, and say that 'Indigenous women in negotiations about mineral development is a fertile area for further research' (ibid.). Kambel (2004: 1) refers to cases that 'document the exclusion of Indigenous women from negotiations and decision making processes relating to their lands and territories, because of erroneous assumptions that this was a man's task.' In the Canadian context, Hipwell et al. (2002: 12) say that male-dominated band councils have usurped the traditional power of women with the result that 'the voices of women are often marginalised' in negotiations. In Indonesia, according to Simatauw (2002: 38–9) women have been marginalised in negotiations with PT Freeport, and denied the right to receive or manage payments made by the company, with the result that 'payments are inclined to benefit men and reduce the roles of women within the community.'

In discussing negotiations in PNG, Bonnell says:

> To date in PNG, all negotiations and decisions regarding mining development have been made by men. In the recently developed forum process [for approving new mining projects], agreements were made between the mine developer, the provincial national and local level governments and the local landowners association. They are all men (Bonnell 1998: 3).

Martha Macintyre, also writing about PNG, states: 'Despite women's legal right to participate in this process [setting the nature of mining arrangements], their voices are rarely heard and they exert very little influence on the miners, politicians and governments who make the decisions about mining projects' (2002: 26). Speaking specifically about the Lihir gold mine, she says: 'Lihirian women played almost no role in negotiations over the mining lease' (2003: 123), and that 'women were excluded from the formal negotiation process. Although they were occasionally consulted by the company's community relations department, women were not represented on the relevant committees and were forced to rely on men to represent their interests' (2002: 27). Byford (2002: 30–1) states that Misima women were excluded from mining negotiations and that as a result 'royalties and compensation payments were invariably paid to the men involved in the negotiations.' Wardlow, discussing gold mining in the PNG highlands, states that 'women do not act as landowner representatives in interactions with the company' (2004: 56), although elsewhere in the same article (ibid.: 63) she refers to two individual women who 'have been very active' in mining negotiations.

A recent roundtable, held by the International Union for the Conservation of Nature (IUCN) and the International Council on Mining and Metals (ICMM) on indigenous peoples and mining companies, noted one case where a project went ahead where the traditional ownership of land was vested in women 'but women were not included in negotiations and received no direct benefits' (IUCN-ICMM 2008: 10). A 2004 review of the current status of indigenous women by the United Nations Permanent Forum on Indigenous Issues (UNPFII 2004: 2) sets the issue in a broader context, stating: 'Indigenous women around the world … agree on one of their greatest mutual concerns: the negative impact of their exclusion from decision making processes that affect them'. It goes on to discuss the specific case of gold mining on Misima Island in PNG, where it claimed that mining company officials chose to negotiate only with male members of the community, 'even though the women owned the land. The result was that the royalties the company eventually paid went directly to the men, and the women had no say over how the money was spent' (ibid.: 3; see also CERD 2004: 6, 26–7).

Similar views emerge from NGOs and activists who focus on the mining industry. The former Oxfam Mining Ombudsman, Ingrid Macdonald, has stated (2003: 6) that the gender impacts of mining include 'companies entering into negotiations only with men, making women neither party to the negotiations, nor beneficiaries of royalties or compensation payments, even in matrilineal societies where women are the culturally recognised landowners'. The campaign against environmentally destructive gold mining, No Dirty Gold (2008), claims: 'Women are often excluded from community negotiations with mining companies', while Friends of the Earth Europe (2008) refers to 'the exclusion or marginalisation of women from negotiation with mining companies regarding project design, mining royalties and compensation systems.' The International Women and Mining Network (IWMN) (RIMM 2004) has called on companies and governments to 'ensure that there is active participation of local women affected by a mining project in decision making,' and say that they 'must take direction from local women about the appropriate ways for ensuring that their views are heard'.

How accurate is this view of exclusion? In my experience, it certainly oversimplifies the situation. In at least some cases indigenous women have in fact played key roles in mining negotiations. This is not to deny that exclusion does occur, but if we are to support, for instance, the call of the IWMN to ensure that there is active participation by indigenous women, it is essential to examine those cases where women do participate effectively, so as to understand why and how it is possible.

If we begin by focusing on the conduct of formal negotiations between indigenous communities and mining companies, there are instances where women have been central. In Canada, for instance, a number of indigenous women have held the position of chief negotiator. To a greater extent than in Australia, agreement negotiations in Canada tend to be organised around the role of chief negotiator, a single individual who has responsibility for the overall conduct of negotiations, under the direction of the relevant decision-making body (which may be a chief in council, or an entire community, as occurs where agreements are the subject of a plebiscite before they can be signed). A number of women have held this role, including the Inuit chief negotiator for the Voisey's Bay impact and benefit agreements (IBAs), signed between Inuit and Innu groups and Inco (now Vale Inco); and for the environmental agreement signed between the groups and the Newfoundland and Canadian governments. The Voisey's Bay agreements are certainly among the most significant signed by indigenous communities in recent years, both because of the scale of economic benefits they bring (revenues of many million dollars a year; and indigenous employment levels among the highest achieved at any large mine in Australia or Canada), but also because of their innovative and extensive provisions regarding indigenous participation in environmental management. Indigenous women have also acted as chief negotiators for major agreements in the Northwest Territories, where one of the Dene communities there, Lutsel K'e, used women negotiators on its agreements with a number of diamond mining companies over the period since 1996 (Bielawski 2003; Weitzner 2006: iii, 12).

Turning to Australia, women played a direct and key role in negotiating the agreement between Argyle Diamonds Ltd and the traditional owners of the Argyle diamond mine in Western Australia from 2002 to 2005. In this case negotiations were conducted by a 'coordinating committee', drawn from all of the native title groups with an interest in the land covered by Argyle's tenements, and supported by a team of logistical staff and technical advisers provided by the Kimberley Land Council (KLC). Key negotiating sessions focused on two or three day meetings between traditional owners, the KLC and the mining company. A majority of coordinating committee members from the key traditional owner groups were women (Argyle Diamonds Limited et al. 2005), and women played a central role in presenting negotiating positions to the company and in pushing for their acceptance. Since 2005 the KLC has supported traditional owners in negotiating three further mining agreements, and in each case agreements have been concluded by negotiating teams that involved substantial participation by women traditional owners.

The negotiations just referred to, it should be stressed, are very substantial in terms of the scale of the projects and investments involved. Also, while Gibson and Kemp (2008) are correct to point out that inclusion of women on negotiating teams does not mean that women are free to contribute equally, in these cases women played central roles, including that of chief negotiator.

Thus the picture of almost total exclusion is oversimplified, at least as it applies to Australia and Canada. This strongly suggests that research focusing on the experience of women involved in these negotiations, on the conditions that made their involvement possible, and on the impact of their participation on negotiation processes and outcomes, would be valuable.

Negotiations: Much More Than Sitting Across a Table

It is also important to look more critically at the concept of 'negotiation' as it applies to mining agreements, in order to fully appreciate the extent to which indigenous women are or are not involved. None of the literature that focuses on the exclusion of indigenous women from negotiations explicitly defines what the process of 'negotiating' a mining agreement actually involves. The general and implicit assumption appears to be that negotiation involves a process of (usually formal) discussion, in which representatives of the parties (mining companies, affected indigenous groups and in some cases government) exchange positions and, over time, reach an agreement that represents the end point of the negotiation. It can be argued that negotiation in fact involves much more than this.

First, and critically, agendas for negotiations must be set. Each party must decide what it will seek to extract from the negotiations, and what it is prepared to concede in order to extract it; in other words what issues to 'put on the table' and how hard to fight for them. As the process of negotiation unfolds, additional, often difficult, decisions have to be made about which issues are vital and on which concessions cannot be made; and which issues can be the subject of compromise. In the wider context of policy making, it has long been recognised that the capacity to control or influence the agenda for decision-making, to determine which issues are 'on the table' and which are not, is critical; and indeed in many cases more important than the capacity to determine outcomes in relation to the issues that do make it on to the table (Bachrach and Baratz

1963; Lukes 1974). The same applies in negotiations. Being a negotiator confers little power if the matters you regard as most important are not on the agenda, or are 'sacrificed' to achieve gains in other areas. On the other hand being able to set the agenda for negotiations confers considerable influence, even if one never sits at a table playing the role of 'negotiator'.

Indigenous women have played a critical role in setting the agenda for a number of major negotiations in Australia and Canada. One example involves negotiations in relation to Comalco's bauxite mining operations at Weipa in Western Cape York, Queensland. By the mid-1990s Comalco had been mining at Weipa for nearly 40 years without the consent or support of aboriginal traditional owners. It initiated negotiations for an agreement in 1996, in part because of policy changes by its parent company, Rio Tinto; and in part because of planned expansions to its mining operations that, though legally sanctioned, might be threatened by the absence of an agreement. An extensive community consultation exercise was undertaken with the support of the regional aboriginal organisation, the Cape York Land Council (CYLC), and under the direction of three steering committees representing the aboriginal communities affected by Comalco's operations (O'Faircheallaigh 2000, 2005). Each committee included a substantial number of women, and in two they outnumbered men by three to two. The committees oversaw the consultations and preparation of economic and social impact assessments that documented the history of Comalco's presence on Cape York and the aspirations and concerns of community members in relation to future mining operations. On this basis a negotiating position was developed and approved by the three committees and, subsequently, by large community meetings.

Women played a critical role in setting the agenda for negotiations. For instance, as a result of their input, the negotiating position emphasised heavily what they characterised as 'recognition and respect'. In their view, much of the negative impact of mining on their community resulted from a fundamental failure by Comalco to afford them recognition as traditional owners. An agreement would have to include a series of measures designed to change this situation, including Comalco's adoption of a corporate policy explicitly affording aboriginal people that recognition, cross cultural awareness training for all Comalco employees, dissemination of information on land ownership in the non-aboriginal community, support for cultural activities and renaming places in the vicinity of the mine with their correct aboriginal names.

The steering committees continued to play a central role as the negotiations developed. For instance, both the communities and the mining company were agreed that provisions governing the use of royalties paid to traditional owners should be determined before an agreement was signed, in order to avoid community conflicts that might develop if no arrangement was in place for royalty distribution when payments commenced. In a series of steering group and community meetings women played a key role in pushing for, and achieving, a trust structure under which a substantial proportion of revenues would be allocated to a long term investment fund, creating a sustainable source of benefit for the communities that could outlast bauxite mining.

For much of the time the group which engaged in face-to-face 'negotiations' with Comalco did not include any indigenous women (although senior women traditional owners did become involved at critical junctures). But to focus on this group and so decide that indigenous women were 'excluded' from the negotiations would be misleading. Aboriginal women played critical roles in determining how a negotiating position would be developed, in shaping the content of that position and in deciding how the financial benefits flowing under an agreement would be utilised. In this way they had a large impact on the eventual outcomes of negotiations (for a summary of the agreement, see Cape York Land Council and Comalco 2001; O'Faircheallaigh 2005).

Women also played a key role in shaping agendas for the Voisey's Bay negotiations, as mentioned earlier. Voisey's Bay is located in the Canadian province of Labrador where the territories of two distinct groups of aboriginal people overlap. The Labrador Inuit are the most southerly of the Inuit nations, while the Innu are the most easterly of the nation of the Algonquin language group. Represented by the Labrador Inuit Association (LIA) and the Innu organisation, Innu Nation, respectively, each conducted separate negotiations and concluded separate agreements with Voisey's Bay Nickel Company (VBNC), a subsidiary of Inco Ltd. Nevertheless, they worked together closely, sharing information and jointly undertaking direct action and litigation to support their negotiating effort.

Prior to undertaking negotiations, Innu Nation established an Innu Nation Task Force on Mining Activities (INTFMA), comprised of three representatives from each of the two Innu communities, 'to provide input into the Innu's decision-making processes regarding [Voisey's Bay]' (INTFMA 1996: 4). The Task Force conducted an extensive process of information dissemination and community consultation, designed to establish Innu aspirations and concerns in relation to mining in general and the Voisey's Bay project in particular. It undertook numerous one-on-one interviews with community members, speaking to approximately equal numbers of men and women (ibid.: 10), and organised small group discussions, information sessions and open community meetings.

The Task Force consultations revealed widespread and deeply felt concerns about the damage mining might cause to land, water and wildlife; and about the negative effects of such damage on the Innu's ability to practice and sustain their culture. Innu had little faith in the willingness or ability of the company or the government to prevent environmental damage, and believed that only their own involvement in decisions about the project could protect the land and its resources. Concern was also raised about damage to Innu cultural heritage, especially burial sites; about the inability of Innu to access the Voisey's Bay area; and about the impact on wildlife of hunting by mineworkers. The prospect of additional jobs and business opportunities was widely welcomed. However, many respondents believed that most such opportunities would accrue to people from outside the region and that jobs obtained by Innu would generally be menial and poorly paid. Concerns were raised that the pace of project development was too rapid to allow appropriate Innu involvement in decision-making, and that the absence of sufficient lead time would make it impossible to train Innu for more highly skilled and better paid positions (INTFMA 1996). The Task Force produced a comprehensive Final Report detailing the benefits and costs community members believed would be associated with mining, and identifying strategies for mitigating negative impacts and maximising positive ones (ibid.).

The LIA undertook a variety of consultation processes to establish its members' goals and priorities in relation to the IBAs. These included Inuit submissions and presentations to the Environmental Review Panel for Voisey's Bay, and the activities of the Tongamiut Inuit Annait (TIA), the organisation representing Inuit women in northern Labrador. The TIA convened a number of workshops on mining agreements, on the environmental review process and on Canada's comprehensive land claims policy. It lobbied to ensure that these processes and policies would include a focus on issues of key interest to Inuit women; and more specifically, it lobbied the LIA on the need for gender equality provisions in mining agreements with Inco/VBNC, and had gender equality provisions drafted for the LIA to include in its negotiations with the company. TIA was informed that Inco rejected these provisions. Archibald and Crnkovich (1999: 28) report that the TIA was not told the reason for the rejection, and that it was unclear whether there were opportunities to negotiate alternative wording or whether some of the provisions could have been achieved at the expense of others. They argue that 'women's representatives, if present, could have advanced the arguments on their own behalf'. In fact, however, gender equality provisions are included in the Voisey's Bay agreements, so it would appear that the Innu and Inuit negotiators refused to concede to Inco/VBNC's rejection of them. This highlights the fact that a capacity to influence negotiation agendas and priorities is extremely important. It supports a key argument of the paper, that a complete assessment of women's role in negotiations must include their influence on these agendas and not just whether they are present at or absent from the negotiating table. Having said that, it must also be remembered that the Inuit chief negotiator was a woman!

Many of the aspirations and concerns arising from the Inuit consultations were identical to those identified by the Innu Nation Task Force (see, for example, Archibald and Crnkovich 1999: 14). These consultative processes represented a critical foundation for Innu and Inuit negotiators in their discussions with Inco/VBNC and with Canada and Newfoundland. This is evident from the content of the (separate) IBAs they signed with Inco/VBNC, and the Environmental Management Agreement they jointly signed with the governments of Newfoundland and Canada. In combination, these comprise a comprehensive package of measures designed to address the concerns and realise the aspirations expressed by Innu and Inuit women and men (Lowe 1998; Innes 2001; LIA 2004: 17–8; Gibson 2006; O'Faircheallaigh 2006b: 49–57, 175–95).

They include a provision, unique in my experience in indigenous agreements with mining companies, under which Inco/VBNC agreed to develop Voisey's Bay at a scale much smaller than it originally intended. This provision substantially reduced its environmental impact and allowed Innu and Inuit people a much longer period over which to develop the capacity to take full advantage of employment and business opportunities. They also provide for Innu and Inuit involvement in government environmental permitting procedures for Voisey's Bay, and a direct role for Innu and Inuit in monitoring the impact of Inco's mining operations. In addition, they contain extensive and detailed provisions designed to enhance employment opportunities. The latter include several commitments by the company: to promote equality of Innu and Inuit men and women; to work with the LIA and the Innu Nation to identify and remove any barriers to the employment and advancement of Inuit and Innu, particularly Innu and Inuit women; to provide gender sensitivity training for the project workforce; and to report on a quarterly basis on the number of Innu/Inuit women hired. The agreements guarantee access for Innu and Inuit to the Voisey's Bay site for hunting and cultural activities, and provide for the protection of Innu and Inuit heritage resources. They also provide for rotation schedules, a cultural leave policy and job-sharing, to facilitate the practice of traditional activities; and prohibit workers from bringing firearms on site and from hunting, fishing or trapping during their employment rotation.

The Argyle negotiations provide a third example. Here women did, as mentioned earlier, play a direct role in negotiations, but in addition were very influential in setting agendas, and also in deciding where concessions would or would not be made in negotiations. The coordinating committee shaped the agenda for negotiations by developing, with the assistance of their support staff, what they called 'traditional owner rules' on each key issue they wished to negotiate about. In some cases these were only half a page in length, but these rules set out core positions in a form that enabled people to refer to them easily, and quickly assess company positions against them. In particular, one issue pushed by women members of the steering committee related to the protection of significant sites. To appreciate this issue, some background on the Argyle project is required.

The Argyle diamond deposit was discovered in 1979, on an escarpment that held a site of considerable spiritual and cultural significance, especially for women. The site was registered with the Western Australian Museum, and the Argyle traditional owners sought to use this fact to secure its protection from mining. However, under Section 18 of the Western Australia's *Aboriginal Heritage Act 1972*, a mining company could apply to the relevant government minister to 'vary the use' of the site, which in the context of mining meant to damage or destroy it. The mining company did this, and received ministerial approval to mine on the site. As the Argyle mine was developed, the location of 'barramundi dreaming' site was destroyed. Its destruction resulted not just in great anguish to its traditional owners, but also in serious social tensions and major repercussions for the women who were the site's primary custodians (Dixon and Dillon 1990).

Given this history, coordination committee members were determined to use the opportunity offered by the negotiations with Argyle Diamonds Ltd to ensure that there would be no repeat of the episode on other parts of the company's lease. They therefore included in the 'rules for cultural heritage' a rule they referred to as 'No means No'. In other words, they sought an undertaking from the company that if traditional owners took the view that an area should not be explored or mined, the company would accept this position and would not apply for ministerial permission to damage the site under the *Aboriginal Heritage Act*. Women members of the coordinating committee fought hard for acceptance of this provision in negotiations, and, when it proved necessary to do so, made decisions about what concession should be made in order to achieve the 'No means No' rules. They were successful, and the Argyle agreement became the first in Western Australia to include such a provision (Argyle Diamonds Limited et al. 2005). Since then a number of other traditional owner groups in Western Australia have achieved a similar outcome in negotiations (KLC 2008).

Post-Agreement Negotiations

These examples illustrate the central importance of the agenda-setting processes that precede and set the framework for formal negotiations leading to the signing of an agreement. Equally, it is important to recognise that negotiations between mining companies and indigenous groups do not stop when agreements are concluded. In addition, structures established as a result of agreements help to determine the allocation of benefits and costs from agreements and more broadly from the projects to which they relate. Both the operation of these structures and ongoing negotiations with companies create further possibilities for the inclusion and exclusion of women.

Agreements create a framework for interactions between indigenous people and companies in relation to mining and related activity on indigenous land. Parts of this framework may be specific and leave little room for subsequent adaptation or interpretation; for example, in the case of financial payments mandated by an agreement. But in many other areas, an ongoing process of negotiation must occur as agreements are implemented and modified, either formally though legal amendments or, much more commonly, through the mutual acceptance of policies and practices that are seen to generate better outcomes for both parties (O'Faircheallaigh 2002; Crooke et al. 2006; for specific examples of this process unfolding, see Gelganyen Trust, Kilkayi Trust 2006: 13–5; Rio Tinto 2008: 20-2, 38–40).

For instance, provisions of agreements designed to provide indigenous training and employment opportunities may set broad goals, indicate a range of initiatives that may be taken to pursue these goals and include an overall commitment of resources by the company. But which specific initiatives should have priority? What should be done if a particular initiative mandated by an agreement proves ineffective? Rarely will either party wish to reopen the agreement to provide for a different approach. This is much more likely to be negotiated informally by company employees and traditional owners 'on the ground'. Regimes established to ensure the protection of aboriginal sites provide another example. The specific requirements contained in an agreement may prove less than ideal for either or both parties, and alternative and more practical alternatives will be agreed. Environmental management provisions in agreements often provide that traditional owners will be consulted about major project changes or about proposed amendments to environmental regimes, and that mine operators will take account of their views (O'Faircheallaigh and Corbett 2005). But how are they to be consulted? How are their views taken into account? To what extent will companies modify their plans? All of these are matters for negotiation.

A full assessment of the role of women in negotiating agreements must also focus on this area. To date it has almost entirely escaped the attention of researchers, but it is clear that in Australia and Canada their role is far from negligible, and indeed in some cases it is central. The issues just discussed are usually handled by joint company—indigenous 'coordination committees' or 'implementation committees' established under agreements, and women often play a key role in these committees. This has certainly been the case with the committees for the Argyle and Comalco Agreements, with both having substantial numbers of women members and having been chaired by aboriginal women. In addition to implementation committees, the Voisey's Bay IBAs provide for the appointment of full time implementation coordinators, and a number of these have been indigenous women, including the current coordinator for the Inuit IBA.

An anecdote, not related to any of the negotiations discussed elsewhere in the chapter, will further illustrate the role of women in this area, and their keen appreciation that negotiations do not end with the signing of an agreement. It involves a major new project in north Australia being developed by a multinational mining company. Negotiations proceeded smoothly, reflecting both the urgency the company faced in reaching an agreement and the fact that the senior company manager responsible for negotiations, and who would also oversee the project's construction, quickly established a reputation for integrity and straight talking. The agreement signing ceremony was in full swing on a beach near the project site when it was noticed that the senior manager and some of the women elders had disappeared. It transpired that they had taken him to a spring further along the beach, and given him a traditional baptism and an aboriginal name. During the period that followed, senior women often had to engage the manager about practical issues related to implementing the agreement. Each time they met him they would greet him by his aboriginal name, and this clearly affected the way he dealt with them. The toughness evident in his dealings with project contractors, union officials and male indigenous leaders was not nearly as obvious, and there always seemed to be a way in which the needs of the women could be accommodated. I am not suggesting that the women's action was cynical; rather it was seen by them as an appropriate way to mark a milestone in their relationship with the manager. But this is precisely the point. The agreement was just a marker; relationships would continue to be negotiated, and they had created a firm foundation on which to ensure that the results of the negotiation would be mutually beneficial.

Examples of structures established under agreements that help determine the allocation of benefits and costs include the trusts often established to manage royalty payments made by project operators, and the environmental monitoring agencies established under a number of Canadian agreements. The trusts created pursuant to the Argyle, Comalco and Voisey's Bay agreements handle millions of dollars in revenues a year (A$6 million in Argyle's case in 2007, for example, and tens of millions of Canadian dollars in 2008 for the Innu and Inuit Voisey's Bay trusts), investing a proportion in long-term capital funds and allocating the remainder to social and economic development in affected communities. All three have had substantial numbers of women trustees, four out of nine in Argyle's case during 2005–2008. The Argyle trusts initially had a woman as deputy chair, and now both co-chairs of the trusts are women. In the case of the Comalco trust, women have served as chairs both of the main trust and of regional trusts that manage the distribution of benefits at the community level.

Structures of this sort are critical in determining whether women share in the benefits of agreements. For instance, the Argyle traditional owners decided to use some of the funds flowing under the agreement to establish a 'Law and Culture' capital fund. The fund is invested in low to medium risk investments to ensure a steady income, and its capital must be preserved. The capital fund, and the income from it, is split evenly between men and women to support their respective cultural activities, and income distributed from it can only be used for this purpose (Gelganyen Trust, Kilkayi Trust 2005: 13).

An example of an environmental agency is the Environmental Management Advisory Board (EMAB), established pursuant to an environmental agreement between Diavik Diamond Mines Incorporated (DDMI), a subsidiary of Rio Tinto and operator of the Diavik diamond mine; five aboriginal groups; and the governments of Canada and of the Northwest Territories. Among the agreement's purposes are to ensure that environmental mitigation measures are appropriately implemented; to undertake additional monitoring to verify the accuracy of environmental assessment and the effectiveness of mitigation measures; and to facilitate the effective participation of aboriginal peoples in achieving these purposes. EMAB's roles include serving as a public watchdog over the environmental regulatory process, and implementing the environmental agreement. It consists of one representative nominated individually by each of the five aboriginal groups, governments of Canada and the Northwest Territories and DDMI.

Since its establishment in 2002 the Board has, for instance, found fault or raised concerns about the way in which Diavik notified regulators of changes to its dike design; about its wildlife monitoring and aquatic effects monitoring programs; its contribution to cumulative effects monitoring; and what the Board saw as DDMI's uncoordinated approach to site visits by aboriginal elders. The Board has also been critical of government regulators, for their perceived failure to maintain adequate inspection of Diavik's operations; to review documents in a timely manner; and to provide sufficient opportunity for the aboriginal communities to comment on license applications. Thus, the Board plays an important role in seeking to ensure that the environmental agreement helps minimise negative environmental impacts from mining, and provide aboriginal people with a real say in environmental management.[2]

In this area also, indigenous women have been active. For example, the Board has drawn on the advice of women elders in developing proposals on how to approach environmental and wildlife management issues, and it has involved

---

2   For details on the Board's roles and activities, see O'Faircheallaigh (2006b: 29–34).

them in environmental monitoring activities. A number of aboriginal community representatives on the Board have been women, and the Board's current Vice Chair is a Lutsel K'e woman, Florence Catholique (EMAB 2004, 2007).

Explaining Opportunities for Participation

While women may be excluded from negotiations in some situations, this is not always the case. What explains these divergent outcomes?

Cultural context is obviously important. For instance, in some regions of PNG, where much of the specific evidence for the exclusion of women in the current literature originates, cultural factors inhibit women's participation. As Macintyre (2002: 28) says: 'The exclusion of women from all-important decision-making … is almost exclusively due to the weight given to Papua New Guinean men's views on "tradition" and the customary role of women.… Mining companies are reluctant to champion women and risk offending "custom"'. In relation to historical traditions on Lihir Island, 'The ritual and political centre of each village was the men's house. There is no evidence that in the past women had any more control over land or ritual events than they do today. Old women report that … male authority [was] less contested than it is now…' (Macintyre 2003: 120).

In contrast, traditionally in many parts of Australia, both men and women had central but separate roles in 'public' spheres related to cultural and religious practices. This separation, while giving 'women's business' particular focuses that may be different from men, 'nevertheless operates in an understanding of the public good in which women have particular and crucial responsibilities' (Edmunds 1996: 127). In Hamilton's words, 'men hold and care for one segment of mythology, and stress particular themes and events in their myths; women hold and care for a complementary set' (cited in Edmunds 1996: 126). This situation certainly applies, for instance, in the Kimberley region where the Argyle diamond mine is located (Rose 1996: 36–7). In the contemporary context, while men may be more inclined to take a public stance, 'men and women continue to maintain their separate but complementary spheres' (Edmunds 1996: 131; see also Rose 1996, 2001; Langton 1998; Kopasur 2002: 14). This situation clearly provides a much stronger basis on which women can exert a major influence on mining negotiations, and it is a situation that is certainly not unique to Australia (Brody 2000: 265–7; Carino 2002: 16).

A second critical issue involves decision-making processes in relation to mining negotiations. One matter involves the openness of those processes and the opportunities they offer for participation. On the indigenous side, the process leading to decisions about what issues to pursue, and how hard to pursue them, can occur through an extensive process of community consultation, deliberation and endorsement that generates a negotiating position which is presented to the mining company. At the other end of the spectrum, it can involve a small group of indigenous people, or even a single person, deciding unilaterally what position to pursue in negotiations. The way in which 'hard' decisions are taken as a negotiation proceeds can similarly vary. To the extent that processes are 'open', the participation of women will be facilitated. In each of the cited examples where women played a key role in setting negotiation agendas and generally in decision-making, the processes involved provided extensive opportunity for community participation and also, in the Australian cases, the establishment of steering committees or similar bodies with broad representation that oversaw the conduct of the negotiations.

The processes discussed share two other important characteristics related to the issue of participation. The first is time. In each case substantial amounts of time were available in which to prepare for negotiations. In the Argyle and Comalco negotiations, this was in part because they involved existing operations and so the pressure of tight project development schedules was absent. In the Voisey's Bay case, the aboriginal groups refused to negotiate until they had undertaken their preparations in the way they wished. The second factor involves resources. Again, in these three cases substantial budgets were available, either from the companies involved or government or a combination of both, to support broadly based consultations and the operation of representative structures. These circumstances applied, in part, because in every case aboriginal political organisations existed (the CYLC, the KLC, the LIA and the Innu nation) that could insist on appropriate time frames and resources and then apply those resources effectively.

Mining company policies can also play a role, and a number of authors highlight the tendency of company representatives to focus on men, and their unwillingness to challenge cultural practices that exclude women (Byford 2002: 30–1; Macintyre 2003: 123). However, their role is secondary in that they tend to adapt their policies to prevailing legal, policy and institutional contexts (Simatauw 2002: 36; Macintyre 2002: 29). This is illustrated by the fact that the same companies have adopted quite different approaches to agreement making in different situations (O'Faircheallaigh 2008: 45).

No one of the factors just discussed is sufficient to guarantee substantial participation by women, and neither does the absence of any one factor rule it out. For example, in negotiations surrounding the first diamond mine in Canada's Northwest Territories, in a context where there was no clear legal requirement for BHP to negotiate, the federal government imposed a 60 day time limit on negotiations. This move, combined with a paucity of resources, created enormous problems for some of the aboriginal communities involved.

Yet indigenous women still played a major role in the negotiations (Bielawski 2003; Weitzner 2006). Conversely, according to Byford (2002: 30) Misima women have been excluded from mining negotiations despite the fact that, historically, they controlled key economic resources and have been able to 'assert their status independently of men'. Overall, however, it appears that the presence of these factors enhances participation by women, while their absence militates against it.

These same factors work to determine the extent to which indigenous communities as a whole benefit from mining negotiations. Research on the extent to which agreements reflect indigenous interests shows clearly that, in addition to the prevailing legal context, critical influences are the presence of a strong indigenous political organisation; the mobilisation of a united community to support the negotiation effort; and the resources that the organisations can marshal and apply, including to allow community mobilisation (Gibson 2006; O'Faircheallaigh 2006a, 2008). Thus, the factors that facilitate participation of women in negotiations are precisely those that generate positive outcomes for communities from negotiations. This highlights an important point. To the extent that communities take the steps necessary to achieve positive outcomes, they will facilitate the participation of women. To the extent that they act to facilitate the participation of women, they will also help to secure positive outcomes for the community as a whole.

Conclusion

There has been little research focusing specifically on the participation of indigenous women in negotiating mining agreements, and there is obviously a need for additional work in this area (Weitzner 2002: 66; RIMM 2004; Gibson and Kemp 2008). That work must be informed by an understanding that much more is involved in negotiating agreements than sitting across a table from mining company representatives. A focus on the interaction between negotiators from each side is appropriate, but it is not sufficient. It is also essential to focus on the processes that set negotiation agendas and determine priorities as negotiation proceed; on the ongoing negotiations that inevitably continue after an agreement is signed; and on the operations of organisational structures established as a result of negotiated agreements. All of these play critical roles in determining the allocation of benefits and costs from agreements and projects and their impacts on women, and so each deserves careful attention.

Particularly when 'negotiations' are seen in this broader context it is clear that, at least in Australia and Canada, indigenous women are not generally excluded from negotiations. The examples cited here do only involve a limited number of projects, but they are highly significant in terms of the wealth they generate and in establishing benchmarks in achieving recognition for aboriginal landowners, and ensuring that negative cultural and environmental impacts are minimised (see, for example, Crooke et al. 2006: 111). In the cases discussed women have not been excluded or marginalised. This does not mean that they have always participated on a basis of complete equality with men, but they have played key and in some cases dominant roles at the negotiating table, in setting negotiation agendas and in the operation of structures established under agreements. Their participation is reflected in the content of agreements and in the distribution of benefits from them, for instance in the inclusion of gender equity provisions in the Voisey's Bay agreements; in the emphasis on traditional owner control over sites (the 'No means No' rule) in the Argyle agreement; in the equal distribution between men and women of income from the Argyle cultural trust; in the emphasis on recognition and respect for traditional owners in the Comalco agreement; and in the allocation of a large part of revenue under the Comalco agreement to creating a sustainable capital fund. It is not, it should be stressed, that aboriginal men opposed these initiatives, but the presence of women and the pressure they generated was critical to their adoption.

The general circumstances that have created the opportunity for women to play a forceful role in negotiations include the existence of legal structures creating opportunities for indigenous groups to negotiate; and specific cultural contexts that are conducive to, or at least do not prohibit, the participation of women. Other vital and more specific factors include the use of processes to prepare for negotiation that emphasise broad community participation and political mobilisation; the availability of sufficient time and resources to ensure that opportunities for participation can actually be realised by women; and the existence of indigenous organisations capable of insisting on appropriate timelines and applying resources effectively. Where these factors operate, they work not only to facilitate women's participation in negotiations but also to enhance the benefits that indigenous communities as a whole gain from them.

References

AIATSIS (Australian Institute of Aboriginal and Torres Strait Islander Studies), 2008. 'Native Title Payments & Benefits: Literature Review.' Canberra: AIATSIS.

Archibald, L. and M. Crnkovich, 1999. *If Gender Mattered: A Case Study of Inuit Women, Land Claims and the Voisey's Bay Nickel Project*. Ottawa: Status of Women Canada.

Argyle Diamonds Limited, Traditional Owners and Kimberley Land Council Aboriginal Corporation, 2005. 'Argyle Diamond Mine Participation Agreement—Indigenous Land Use Agreement.' Perth.

Bachrach, P. and M.S. Baratz, 1963. 'Decisions and Non-decisions.' *American Political Science Review* 57: 641–51.

Bielawski, E., 2003. *Rogue Diamonds: Northern Riches on Dene Land*. Vancouver: Douglas and McIntyre.

Bonnell, S., 1998. 'Impact of Mining on Women.' Paper presented at conference on 'Papua New Guinea Mining and the Community', Madang, 26-29 July.

Brody, H., 2000. *The Other Side of Eden: Hunters, Farmers and the Shaping of the World*. Vancouver: Douglas and McIntyre.

Byford, J., 2002. 'One Day Rich: Community Perceptions of the Impact of the Placer Dome Gold Mine, Misima Island, Papua New Guinea.' In I. Macdonald and C. Rowland (eds), op.cit.

Cape York Land Council and Comalco, 2001. 'A Way Forward Together.' Press Release, 11 March.

Carino, J.K., 2002. 'Women and Mining in the Cordillera and the International Women and Mining Network.' In I. Macdonald and C. Rowland (eds), op. cit.

CERD (Centre for Environmental Research and Development), 2004. 'Echoes in the Wilderness: Mining, Women and Communities Workshop August 2004' Boroko: CERD.

Crooke, P., B. Harvey and L. Langton, 2006. 'Implementing and Monitoring Indigenous Land Use Agreements in the Minerals Industry: The Western Cape Communities Co-existence Agreement.' In M. Langton, L. Palmer, K. Shain and O. Mazel (eds), *Settling with Indigenous People: Modern Treaty and Agreement Making*. Sydney: The Federation Press.

Dixon, R. and M. Dillon (eds), 1990. *Aborigines and Diamond Mining: The Politics of Resource Development in the East Kimberley*. Nedlands: University of Western Australia Press.

Edmunds, M., 1996. 'Redefining Place: Aboriginal Women and Change.' In R. Howitt, J. Connell and P. Hirsch (eds), *Resources, Nations and Indigenous Peoples: Case Studies from Australasia, Melanesia and Southeast Asia*. Melbourne: Oxford University Press.

EMAB (Environmental Monitoring Advisory Board), 2004. 'Annual Report 2003/2004.' Yellowknife: EMAB.

————, 2007. 'Annual Report 2006/2007.' Yellowknife: EMAB.

Flanagan, F., 2002. 'Pastoral Access Protocols: The Corrosion of Native Title by Contract.' AIATSIS, Canberra (Native Title Research Unit Issues Paper 2(19)).

Friends of the Earth Europe, 2008. 'Extractive Industries: Blessing or Curse?' Viewed 17 october 2008 at www.foeeurope.org/

Gelganyen Trust, Kilkayi Trust, 2005. 'Future, Country. Gelganyen Kilkayi.' Kununurra: Gelganyen Trust, Kilkayi Trust.

————, 2006. 'Gelganyen Trust, Kilkayi Trust: Annual Report April 2005–June 2006.' Kununurra: Gelganyen Trust, Kilkayi Trust.

Gibson, R.B., 2006. 'Sustainability Assessment and Conflict Resolution: Reaching Agreement to Proceed with the Voisey's bay Nickel Mine.' *Journal of Cleaner Production* 14(3–4): 334–8.

Gibson, G. and D. Kemp, 2008. 'Corporate Engagement with Indigenous Women in the Minerals Industry: Making Space for Theory.' In C. O'Faircheallaigh and S. Ali (eds), *Earth Matters: Indigenous Peoples, the Extractive Industries and Corporate Social Responsibility*. Sheffield: Greenleaf Publishing.

Hipwell, W., K. Mamen, V. Weitzner and G. Whiteman, 2002. *Aboriginal Peoples and Mining in Canada: Consultation, Participation and Prospects for Change*. Ottawa: North South Institute.

Innes, L., 2001. 'Staking Claims: Innu Rights and Mining Claims at Voisey's Bay.' *Cultural Survival Quarterly* 25(1): 12–16.

INTFMA (Innu Nation Task Force on Mining Activities), 1996. *Ntesinan Nteshiniminan Nteniunan: Between a Rock and a Hard Place*. Sheshatshiu: Innu Nation.

IUCN-ICMM (International Union for the Conservation of Nature and the International Council on Mining and Metals), 2008. 'Mining and Indigenous Peoples Issues Roundtable: Continuing a Dialogue between Indigenous Peoples and Mining Companies.' Viewed on 17 October 2008 at http://www. icmm.com/document/237

Kambel, E.R., 2004. *A Guide to Indigenous Women's Rights under the International Convention on the Elimination of All Forms of Discrimination against Women*. Forest Peoples Programme. UK: Moreton-in-Marsh.

KLC (Kimberley Land Council), 2008. 'Completed Agreements.' Viewed 18 March 2008 at http://www.klc.org.au/agrees_complete.htm

Kopasur, P., 2002. 'An Australian Indigenous Women's Perspective: Indigenous life and Mining.' In I. Macdonald and C. Rowland (eds), op. cit.

LIA (Labrador Inuit Association), 2004. 'Labrador Inuit Association Annual Report 2003–2004.' Nain: LIA.

Langton, M., 1998. 'Grandmothers' Law, Company Business and Succession in Changing Aboriginal Land Tenure Systems.' In W.H. Edwards (ed.), *Traditional Aboriginal Society*. Melbourne: Macmillan.

Lowe, M., 1998. *Premature Bonanza: Standoff at Voisey's Bay*. Ontario: Between the Lines.

Lukes, S., 1974. *Power: A Radical View*. London: Macmillan.

Macdonald, I., 2003. 'Tunnel Vision: Women's Rights Undermined.' Paper presented at conference on 'Women in Mining', Madang, 3 August.

Macdonald, I. and C. Rowland (eds), *Tunnel Vision: Women, Mining and Communities*. Melbourne: Oxfam Community Aid Abroad.

Macintyre, M., 2002. 'Women and Mining Projects in Papua New Guinea: Problems of Consultation, Representation and Women's Rights as Citizens.' In I. Macdonald and C. Rowland (eds), op. cit.

———, 2003. 'Petztorme Women: Responding to Change in Lihir, Papua New Guinea.' *Oceania* 74(1–2): 120–31.

No Dirty Gold, 2008. 'Women and Mining.' Viewed 17 October 2008 at www.nodirtygold.org/disadvantaged_women.cfm

O'Faircheallaigh, C., 2000. *The Cape York Model of Project Negotiation*. Canberra: Australian Institute for Aboriginal and Torres Strait Islander Studies.

———, 2002. *A New Model of Policy Evaluation: Mining and Indigenous People*. Aldershot: Ashgate Press.

———, 2004. 'Evaluating Agreements between Indigenous Peoples and Resource Developers.' In M. Langton, M. Tehan, L. Palmer and K. Shain (eds), *Honour Among Nations? Treaties and Agreements with Indigenous People*. Melbourne: Melbourne University Press.

———, 2005. 'Creating Opportunities for Positive Engagement: Aboriginal People, Government and Resource Development in Australia.' Paper presented at conference on 'Engaging Communities', Brisbane, 12–17 August.

————, 2006a. 'Aborigines, Mining Companies and the State in Contemporary Australia: A New Political Economy or "Business as Usual"?' *Australian Journal of Political Science* 41(1): 1–22.

————, 2006b. *Environmental Agreements in Canada: Aboriginal Participation, EIA Follow-Up and Environmental Management of Major Projects.* Calgary: Canadian Institute of Resources Law, University of Calgary.

————, 2008. 'Negotiating Protection of the Sacred? Aboriginal-Mining Company Agreements in Australia.' *Development and Change* 39(1): 25–51.

O'Faircheallaigh, C. and T. Corbett, 2005. 'Indigenous Participation in Environmental Management of Mining Projects: The Role of Negotiated Agreements.' *Environmental Politics* 14(5): 629–47.

Public Policy Forum, 2005. *Sharing the Benefits of Resource Developments: A Study of First Nations—Industry Impact Benefits Agreements.* Ottawa: Public Policy Forum.

RIMM (International Women and Mining Network), 2004. 'Resolution on Indigenous Peoples and Women.' Viewed 17 October 2008 at www.theminingnewsorg/news.cfm?newsID=83

Rio Tinto, 2008. 'Sustainable Development Report Argyle Diamonds 2007.' Viewed 20 October 2008 at www.riotinto.com/documents/ReportsPublications/2007_Argyle_Diamonds_sustainable_development.pdf

Rose, D.B., 1996. *Nourishing Terrains: Australian Aboriginal Views of Landscape and Wilderness.* Canberra: Australian Heritage Commission.

————, 2001. 'The Silence and Power of Women.' In P. Brock (ed.), *Words and Silences: Aboriginal Women, Politics and Land.* Sydney: Allen and Unwin.

Simatauw, M., 2002. 'The Polarisation of the People and the State on the Interests of the Political Economy and Women's Struggle to Defend their Existence: A Critique of Mining Policy in Indonesia.' In I. Macdonald and C. Rowland (eds), op. cit.

UNPFII (United Nations Permanent Forum on Indigenous Issues), 2004. 'Indigenous Women Today: At Risk and a Force for Change.' Viewed 17 October 2008 at www.un.org/hr/indigenousfourm/women/html

Wardlow, H., 2004. 'The Mount Kare Python: Huli Myths and Gendered Fantasies of Agency.' In A. Rumsey and J. Weiner (eds), *Mining and Indigenous Lifeworlds in Australia and Papua New Guinea.* Wantage: Sean Kingston Publishing.

Weitzner, V., 2002. *Through Indigenous Eyes: Towards Appropriate Decision-Making Processes Regarding Mining on or Near Ancestral Lands: Final Synthesis Report Phase 1*. Ottawa: The North-South Institute.

———, 2006. *Dealing Full Force: Lutsel K'e Dene First Nation's Experience Negotiating with Mining Companies*. Ottawa: North South Institute and Lutsel K'e Dene First Nation.

7. Gender-Based Evaluation of Development Projects: The LAST Method

Kuntala Lahiri-Dutt

Introduction: An Alternative Inquiry

This chapter addresses three key research questions often posed in field-based and participatory development research: how to effectively integrate Monitoring and Evaluation (M&E) into the project cycle; how to integrate a gendered approach to participatory surveys; and how to use an assets-based approach as opposed to a conventional needs-based assessment. The innovation in this chapter lies in the use of a recently developed asset-based participatory M&E method (Livelihood Asset Status Tracking or LAST), in combination with gender analysis, undertaken by participants to examine the impacts of community development projects on women and men in a coal mining region of East Kalimantan, Indonesia. The work was undertaken in and around the areas of operation of PT Kaltim Prima Coal (KPC) in Sangatta, East Kalimantan, as part of an Australian Research Council-funded Linkage Project. The Australian National University (ANU) and PT Kaltim Prima Coal are joint project partners.

Two Stories

I will start with two stories that comprise the background for this research. In early 2008, I met a senior Community Relations manager of a large multinational mining company. She had joined this company after serving a number of years with non-government development organisations, and expressed her surprise at how the mining industry has generally neglected simple M&E processes in its community development projects. Such processes are routinely used by development and donor agencies. Indeed, the Community Development Toolkit published by the International Council on Mining and Metals (ICMM) has just one section on M&E, but uses a Logical Framework (logframe) approach, which is based on managerial principles rather than community participation. For example, it uses a matrix of goals, outcomes, outputs and inputs against indicators and their measurement. Our idea was to explore ways in which we could go about the M&E process that would not be essentially top-down, but involve community members so that the process of evaluation itself becomes an awareness raising exercise for both company personnel involved in the daily running of the projects and the community.

The other story is also personal and dates back to 2004, when the then Managing Director of KPC invited me to find out 'what the women in the community want'. It is not unusual for researchers to undertake 'needs assessments',[1] and indeed such assessments of women's and men's needs and interests comprise a most important part of gender analysis for many development interventions. Consequently, I undertook a systematic needs assessment exercise with women in three village clusters around the area of mine operations (see Lahiri-Dutt 2004). I also explored what people think and how they feel about the investments made by the development agent—KPC.

Although these assessments are a good initial step towards promoting understanding, collaboration and partnership between community members and a development agent, they are still located within the 'needs' framework, and do not fully expose the entire livelihood dynamics and the social and economic well-being of women as well as men within communities. This exercise led me to question whether there could be other, better, ways to explore the gendered worlds within mine-affected communities. That exploration led to the development of an action-research project[2] to integrate gender into the planning, monitoring and evaluation of community development projects devised by KPC.

Asset Mapping

Asset mapping—as opposed to needs assessment—is not just a technique; it is an alternative way of thinking about community development, in the sense that it is seen as an open-sum perspective that focuses on the abilities and capacities, rather than the deficiencies, of the community.[3] Whereas conventional community development is often implemented in a 'top down' and 'outside-in' manner, assets-based community development is 'inside-out': the dependence on the outside expert or consultant as a direct implementer of projects is minimal. The emphasis in 'Asset-Based Community-Driven' (ABCD) development processes is on the community's or on an individual's ability to

---

1   Needs assessments are systematic explorations of the way things are for women and the way they should be, and of what people are thinking and how they feel (Barry et al. 2000). Guy (1996) thinks that while information from a needs assessment study is valuable and useful, the process of gathering such information itself is also valuable in itself.

2   See Chapter 1 of this volume for further details of the ANU-KPC, ARC-funded Linkage Project, 'Creating Empowered Communities: Gender and Sustainable Livelihoods in a Coal Mining Region in Indonesia'.

3   Community assets may include the collective talents and skills of community members; existing networks and associations, institutions; physical assets such as land, buildings and equipment; economic assets such as local businesses and exchanges; and cultural assets such as heritage and histories.

play a 'catalytic role' as an awareness-builder as well as a facilitator and learner. The ABCD process is inherently optimistic and one that focuses on strengths rather than the problems and deficits that are identified in needs-based assessments. The problem with this approach is that there is a possibility that the community might internalise some of the deficiency attitudes and come to depend on outside resources or professional experts to find solutions to their problems. Experience tells us that when such a solution is 'found' by others for a community, it generally is not sustainable. Assets-based assessment methods are thus becoming popular for non-government organisations (NGO) and donor-sponsored community empowerment programs in Latin America and Africa.[4]

What is New in this Work?

The challenge in the fieldwork reported here was to combine an assets-based approach with gender-sensitive tools to understand the benefit streams and impacts of development projects run by KPC on men and women in the community. A further challenge was to undertake the research during the project cycle rather than at the end, which is usually when most M&E tools are used, although monitoring is typically undertaken as an ongoing process.

First, our evaluation occurred not after the project was over but during the project cycle, making it essentially a process evaluation rather than an end-evaluation. Second, our intention was to offer new ways of assessing and learning that are more inclusive and more in tune with the views and aspirations of those directly affected by the projects. Finally, and most importantly, a great amount of emphasis was put on differentiating our observations and analysis along gender lines. A pre-condition for achieving people-centred development is the removal of existing disparities between the social, economic and political status of women and men. Our project, therefore, deviated from prior M&E exercises in three key ways by:

1. conducting a 'process evaluation';

2. using participatory M&E; and

3. seeing the M&E process through a gender lens.

4 For example, Goulet (2008) has outlined how a self-conscious framework for understanding how change takes place within a community might be a better approach to initiating change, citing the example of an Ethiopian rural community that become more self-reliant when this approach was taken by the development agent, Oxfam and a local NGO, the Hundee Grassroots Development Initiative.

Sustainable Livelihoods Approach

Our research and fieldwork was undertaken within the overall framework known as the 'sustainable livelihoods' approach. Originally developed by Robert Chambers, of the Institute of Development Studies (IDS) in the UK, to understand the complex experience of vulnerability and access to natural resources, the concepts in this approach have been incorporated into the review and impact assessment of development projects and policies on a multi-dimensional basis (Ashley and Hussein 2001; Nicole 2001; also see FAO and ILO 2009). According to Whitehead (2002: 575), the framework incorporates a sustained critique of externally imposed definitions of poverty, especially the one dimensional income and consumption line approach. The tools used to conduct sustainable livelihoods assessment are still evolving amid significant debate as to their theoretical, methodological and empirical veracity. The lack of quantification in the approach poses a challenge to conventional development planning and makes it difficult to implement in practice, although the general agreement is that the livelihoods approach is useful for a pro-poor focus and holistic analysis of factors that are locally relevant (Turton 2001; Bond and Mukherjee 2002).

As an example, in agricultural research there has been a noticeable move beyond increasing food production to addressing the larger goal of reducing poverty. Adato and Meinzen-Dick (2002) utilised the livelihoods approach for an expanded understanding of poverty that goes beyond income or consumption-based headcounts and severity measures, to consider the many other factors that poor people in different contexts define as contributing to their vulnerability, poverty and well-being. The sustainable livelihoods framework thus provides a common conceptual approach that has the ability to shift the focus towards livelihood outcomes, rather than project objectives, and to the full range of impacts rather than just cash incomes and physical outputs within the beneficiary community.

The philosophy behind the use of a gender lens is that the empowerment of poor people and particularly poor women is the key to long-term poverty alleviation through community empowerment. Women in poor communities in rural Kalimantan, where our project is located, often hold far less power and official recognition as economic agents than men, while being largely responsible for household chores. The ANU-KPC Linkage Project sought to change this situation and was based on the assumption that community development projects must help women build their confidence, increase their self-reliance and assist them in making decisions and maintaining control over their resources within the family and the community.

Women living in the rural areas around KPC's mine operation are and have always been active in supplementing family incomes in a number of ways and play important economic roles not only within the household but also through engaging in trade and commerce. Company staff, however, almost always chose to ignore these productive roles and followed the official Indonesian definition which sees the male as the 'household head'. According to Kabeer (2005) such a conflation of sexual or biological difference with gender or social difference arises from the view that the roles, capacities and aptitudes attributed to men and women within a context are rooted in their biology and hence cannot be changed. In deciding to integrate a gender perspective into KPC's community development policy and programs it was recognised early on that the empowerment of women should not increase their burden of work but increase their part in decision-making affecting their communities. The task was to develop an analytical framework that could translate feminist insights into action and integrate gender perspectives into the community development process. The deeply entrenched and institutionalised nature of gender inequalities within the company's bureaucracy was a challenge that all of us involved in the project also had to overcome.

Monitoring and Evaluation with LAST

M&E, in brief, is a basket of tools to assess whether community development projects have succeeded or failed. M&E is widely used by governments, donor agencies and organisations implementing development projects to provide accountability, indicate to donors the effectiveness of past expenditure, assure them of quality and pinpoint areas of strength and potential weaknesses. While the term 'monitoring' gives the impression of a continuing function—an ongoing supervision during the life of a project or a program—the process of evaluation is a selective exercise to systematically assess progress towards the achievement of an outcome, and is more costly and less frequent than monitoring (UNDP 2002). Koch (1994: 1148) notes: 'Evaluations are usually engaged with measurement, and their results are used by managers to control and predict aspects of the … service.' In general, M&E processes focus on inputs (funds, experts, equipment), outputs (number of people trained), outcomes (increased incomes or creation of jobs), impacts (improved health or longevity) and the objectives and goals of projects. These processes can themselves lead to improved capacity amongst those who undertake them. Figure 7-1 displays some of the objectives of M&E processes.

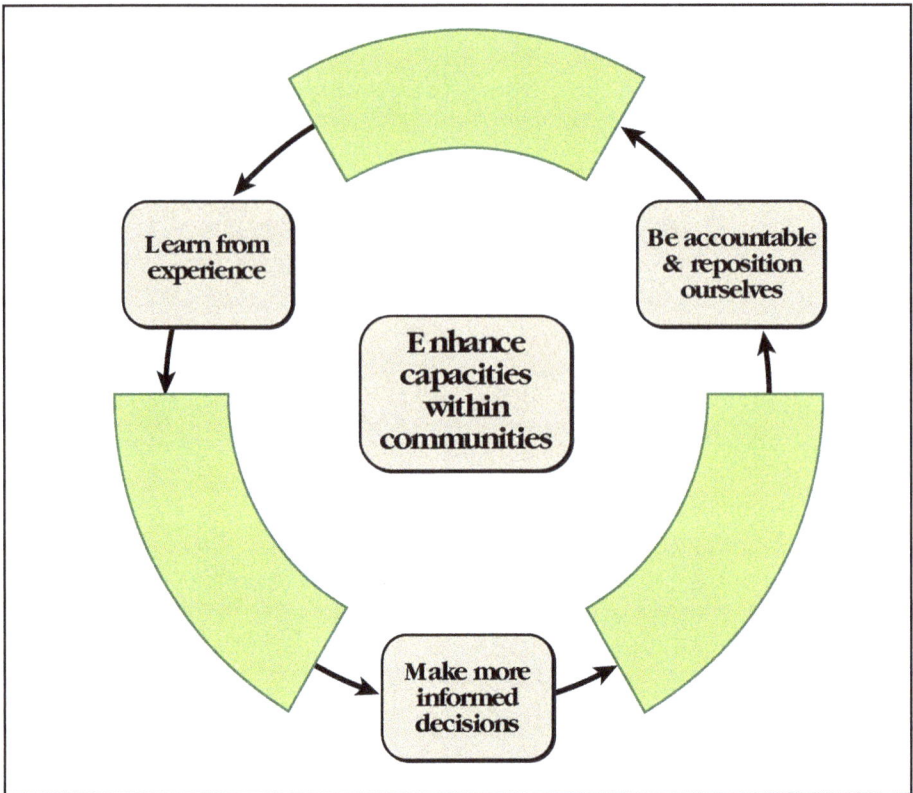

Figure 7-1: Objectives of Monitoring and Evaluation.

Source: Adapted from various sources.

Conventionally, M&E has been undertaken in the development context by outside experts coming in to measure performance against previously set indicators. For this, a range of procedures and tools have been developed. Often these tools—associated with scientific management of development projects— are used uncritically by external experts hired by funding agencies.

Monitoring presents serious challenges for 'process' projects. Instead of just externally assessing (often) sub-optimum impacts, a robust method must internally monitor emerging impacts to enhance the effectiveness of project implementation. For this reason, Bond and Mukherjee (2002) use the term 'impact monitoring' for process evaluation, which is now increasingly demanded by funding agencies and donors. Initially tested in Eastern and Southern Africa as 'stakeholder analysis and local identification of indicators' this method was first used to examine the success and sustainability of farming-based livelihood systems. Such monitoring can detect changes in impact during the implementation of a project. This process was refined by Bond and Mukherjee (2002) into the LAST method, defined by them as:

A rapid impact monitoring system designed along with primary stakeholders and based on the Sustainable Rural Livelihoods conceptual framework. It is intended to track the ongoing dynamics of five capital assets (usually natural, physical, human, financial and social) essential to household livelihoods as a proxy for impact.

The LAST method was implemented in 2000 by the Indian Farm Forestry Development Co-operative in Pratapgarh, Rajasthan and has been used for retrospective impact assessment by the Gramin Vikas Trust of the Eastern India Rainfed Farming Project in Ranchi, Jharkhand.

The Origins of LAST

Both conceptually and practically, the LAST method has several predecessors. Conceptually, it derives from the concept of 'sustainable development' but goes beyond an economic definition of development (the production of goods and services for general utility) to incorporate both a social perspective (access, inclusion and intra-generational equity) and an environmental perspective (conservation of natural capital and inter-generational equity). It also incorporates the 'sustainable livelihoods' framework as the analytical basis for the asset endowment of the rural poor.[5] Essentially it denotes what Robert Chambers (1989) described as a 'new professionalism' within the development field that recognises local autonomy of action and the fostering of innovation within organisations.

Practically, LAST involves methods such as wealth ranking, a Participatory Rural Appraisal (PRA) technique that encourages rural people to classify households in terms of relative wealth or well-being according to their own criteria and judgment. Quality of life indices are used to identify a range of locally meaningful descriptions of household situations from worst to best known, which participants then scale and score. Bond and Hulme used this method as early as 1992 to investigate the operations of partners in rural India. Within the Sustainable Livelihoods framework, the LAST method begins as a participatory assessment of assets to which rural people have access in order to devise their livelihood strategies. According to Bond and Mukherjee (2002: 808) this is also the end point of LAST, 'as those strategies impact not only on their livelihoods in terms of outcomes (a more traditional source of indicators) but also back on the assets themselves.' Thus, the changing asset base, measured for the five types of capital to which a household has access, can be a useful proxy for impact on livelihoods. Two elements are important in this method:

---

5 I am aware that seeing 'assets' as capital-based is contentious, but in this instance will avoid that specific debate.

participation and rapidity. If indicators are derived in a participatory way, they will be locally relevant within relatively homogeneous areas (in terms of ethnicity and biodiversity/agriculture). The method needs to be quick and simple enough for rapid enumeration if it is to be used over many households with a reasonable frequency.

Development projects initiated by external agents, such as the mining company KPC in this case, are notorious for their male bias (Elson 1991). The challenge for us was to track the male bias, identify the roots of such bias and show to all participants how women in rural Kalimantan are dealing with the inherent male bias in development projects by using their limited assets innovatively.

LAST Workshops with Men and Women

For us, the main vehicle for undertaking LAST was a series of participatory workshops conducted with the project team and project beneficiaries of two of KPC's community development projects. The workshops involved small group discussions, brain-storming and clustering of criteria, field-testing and validation. We invited both women and men to participate in these workshops and used a number of cards, slides and posters, flip charts and boards. Initially women were reluctant to participate with men present, but as we began to speak on the assets and strengths of individuals, families and village communities, women expressed more interest in speaking up. Since the workshops were also used as a means of gender sensitisation, we included broad-ranging discussions on who does what in the community (gender roles), how people lived in the past and the present and how much effect the project has had on women's and men's lives within the household (before-after scenarios). As per Bond and Mukherjee (2002: 808), the main objective was to evolve 'word pictures' for constructing verbal descriptions of asset status. Such word pictures depict the 'worst off' and 'better off' households and also intermediate positions. We had two sets of workshops to identify the impacts on the livelihoods of women and men of two community development projects: a Tie and Dye (T&D) Project and an Orange Cultivation (OC) Project. The first is based in villages around an urban area and the second is located in a rural cluster of planned transmigration villages located a considerable distance from the nearby town, and characterised by poor transportation options.

Participant Profile

There were 11 participants in our workshops from the T&D project and 23 from the OC project. As the T&D project involves women community members only, disaggregation by sex was possible only for the OC project, with nine women and 14 men in our workshop. The average age of workshop participants was 34 for women and 42 for men in the OC project, and 41 for women in the T&D project. Being located in Rantau Pulung, which is the more remote of the village clusters, the OC project featured poorer and less literate participants (12 of the 23 participants had a primary school education only), but there was no relationship between literacy levels and incomes. Similarly, there was no correlation between low educational achievement and income among the T&D project participants. As many as six OC participants, however, reported 'uncertain' incomes and the greatest concentration of participants were in the Rp450 000–500 000 per month income category. In comparison, those in T&D project, being located in Sangatta town, reported monthly incomes that had a wider spread: three participants earned less than Rp350 000; three others Rp500 000, while two reported incomes of Rp1 000 000, and one over Rp2 500 000 per month.

The level of home ownership was considerably greater for participants in the OC workshop, with only one participant reporting as currently living in a house he did not own, whereas just under half of the participants in the T&D workshop owned their own home. Of the remainder three reported as living with others and three were renting. These participants had lower income levels (less than Rp400 000 per month). However, there was a positive relationship between home ownership amongst T&D participants and length of stay—it was uncommon for those living in Sangatta for more than six years to rent or stay with others. The patterns were clear: being from a transmigration area, OC participants tended to own a home (given by the government), and the self-motivated migrants living in or around Sangatta initially either rent and/or stay with others while building up a broader income base.

In terms of gender roles and occupations, as many as eight of the women in the OC Workshop self-identified as 'housewife' and only one identified as 'farmer', while 11 of the men self-identified as 'farmer'. Being urban-based, the participants in the T&D project represented a wider variety of occupations— five reported as housewives; two as traders; two as working in small business and two in tailoring/sewing. Of the 23 participants in the OC workshop, 14 reported other sources of income. In our workshops we focused on both the main and secondary sources of income in order to explore the full range of economic activities that the households performed. A large number of farming families in the OC project also reared poultry. One family had cattle, another had a store and one other also depended on hunting. As the T&D project involved

119

women only, the participants reported having husbands who were working in a range of occupations, as truck drivers, casual workers or in small business, but only two female participants were engaged in additional income-generating activities—one in catering and one in sewing.

Sangatta itself is a relatively new urban development in East Kalimantan and so most of the workshop participants came from migrant families. The OC project runs in a transmigration area—19 participants had lived there for around 10 years, two for eight years and one for nine years—revealing a relative uniformity in the length of stay in the area. The participants in the T&D project reported periods of residence in Sangatta ranging from one year to 25 years. Similarly, the OC project had far less ethnic diversity: as many as 12 participants were Javanese and four were from Flores, the remainder included two each from the Timur and Sunda ethnic communities, and one each from Bima, Ciamis Jabar and Nusa Tenggara. By comparison, the largest ethnic community represented amongst the T&D participants was Bugis (three), the rest were made of one person each from Banjar, Banjar Manado, Central Sulawesi, Dayak-Chinese, Kutai Kartanegara, Madura and Toraja.

The Tie and Dye Project

As described, the T&D project is an urban-based, 'women only' initiative. There are 40 members of a women's group (Lembaga Pemberdayaan Wanita) involved in the project under the leadership of Ibu Ariati. The focus group meetings were held in her house, as were the training sessions. The project statement read:

> Women in Sangatta want to be part of an empowering process and social change, which also involves nature and the environment, and want to produce creative art and innovate through local art and crafts.

The project is still in the early stages; it began in May 2006, and at the time we undertook our fieldwork, there had been only two training sessions and one exhibition of the resulting fabric products. The idea behind the project was to continue a local craft tradition using natural resources sourced from the surrounding area. It was envisaged that the project would strengthen local institutions and involve women in them. The raw materials—both the fabric and the threads—are provided by KPC. The company also hired an expert for the project to train the participants to collect, prepare and use natural dyes. The project was dependent on the materials supplied by KPC. Fixing of the natural dyes is done by using chalk and alum. The fixing is generally effective and the colour does not run except during the first wash, although it is not certain how long the dye lasts over the long-term life of the product.

During our focus group discussions, participants expressed a range of concerns about the long-term viability of the initiative. A critical issue for the project is that of financial sustainability—once the initial materials are used up, the villagers may not have the means to continue the work and practice the newly acquired skills. Pak Faddin, a member of the Community Empowerment Department of KPC, had looked at the problem of prices both for buying materials and selling the products. The initial materials provided by KPC were bought at inflated prices because KPC is always given higher than normal prices. Should the women go on to buy their own fabric, they will probably manage to negotiate lower prices, especially if sourced from local towns such as Balikpapan or Samarinda, or even ordered from Jakarta. However, the participants faced another problem in getting the product to market, as the group did not yet have connections to shops where they might sell their products. Many of the participants thought that eventually having a small boutique shop would be a way to solve this problem. The main concern of the focus group participants was ensuring long-term commercial viability and independence from KPC, and they saw the main obstacle to this as a lack of initial capital to buy the necessary basic materials. Concern was also expressed about the market potential of the products created by the group. However, as noted by Ibu Marita, the process of learning a new skill—in only a few hours a day—that proved to be useful in earning additional incomes was an empowering experience. Ibu Marita has a small shop from which she sells beads and other trinkets and hopes to also sell the T&D products one day.

The Orange Crop Project

Orange crops have been planted in Rantau Pulung since the 1990s, when the area was settled through transmigration schemes. The local ecology proved to be ideal for orange cultivation and the shady trees helped to prevent soil erosion. Although a few farmers have started to receive benefits, some have faced problems relating to poor quality harvests which have discouraged local business merchants from purchasing them. Based on these observations and a needs assessment, the OC project was devised as a Community Empowerment program in 2004, primarily to assist farmers to improve their skills and provide them with certified seeds, fertilisers and other technical assistance. The project had a tool supply component in addition to the improved seeds supply, as well as consultations by external experts to provide skills development and training in land preparation, planting, cultivation management and pest control. It also included physical verification by a team comprising agricultural experts, KPC Community Empowerment staff, local government representatives and community members themselves, to eliminate the spread of 'orange virus'. The project also included capacity-building through in-house training and study in East Java. As part of the project, field facilitators were employed to visit farmers and gather data to map their perceptions.

The OC project had 115 beneficiaries—less than the number originally envisaged—due to a limited number of seeds being available. All participants were men because in Kalimantan, the ownership of land is usually vested in a man, although women work in family farms. The recruitment of project participants was done more or less informally in meetings with village communities, based on willingness and level of interest and the ability to provide land and other raw materials required to participate. The productivity of orange cultivation in East Kalimantan is higher than the national average. Marketing of the fruit is the responsibility of farmers themselves and Sangatta is the main market, with a high demand for the fruit. Orange is a farm product with cash potential—a household can expect to earn around Rp4.5 million within three years of planting. Oranges also have the potential to diversify the diets of local people, particularly the poor, and provide improved nutrition and health outcomes.

Our initial observation was that the OC project has been gender-blind in the early conception and recruitment stages. Women were not seen as direct beneficiaries and the project primarily targeted men, who owned the family lands. However, although the main beneficiary was in most cases the male head of household, it was quickly discovered that women were frequently involved in the maintenance of orange plantations, as well as the harvesting of the fruit, and sales. During the workshops, it was revealed that women provide a significant amount of labour both in household gardens and orange cultivation. The extension workers also reported that women were hungry for knowledge and many accompanied the visiting agricultural experts when they toured around the village plots during their inspection trips. Women were most likely to be present in the training sessions which covered fertilisation and other elements of plant care. Once the project commenced, the high interest and involvement of women provided impetus for integrating a gender perspective in the OC project. This lead to a number of changes to the project, such as setting the timing of field visits by external experts and extension workers to suit women, and the replacement of expensive chemical fertilisers in favour of organic manure composted from animal dung and household wastes. Women participants argued that although the use of organic nutrients involved more labour, requiring the assistance of their husbands and/or older family members, they reduced costs drastically, as well as their dependence on the company's supply of fertiliser. The use of organic manure, they commented, also decreases exposure to synthetic chemicals and has beneficial effects on health, in particular the skin on the hands.

One of the main difficulties for the project related to culturally-based assumptions about the make-up of rural agricultural communities around Sangatta. While the settled Javanese transmigrant communities—most of whom were farmers in Java—adapted easily to a farming-based livelihood and thus received the most benefits from the project, local Dayak and Kutai peoples who have traditionally had a forest-based livelihood failed to take full advantage of the project. In envisioning the participants as land-based farmers, the project left out such groups of people—an exclusion which indirectly encouraged them to take up slash and burn agriculture.

In terms of practical obstacles, getting the fruit to market proved difficult, as the villages were located on average about two to three hours from Sangatta. This resulted in the formation of a buyers' market, which the trader from Sangatta visited once or twice a week to collect the oranges, leading to the families receiving a lower price than they could get in town. For women, poor access to the market and the lack of transportation facilities were the two biggest hurdles which prevented them from taking full advantage of the program.

In conversations with some of the participants of the OC project, it emerged that many felt that the project had positively contributed to the well-being of farming families. Improvements in the quality of life had not only come about from access to greater incomes, but as one participant observed, a 'sense of self-reliance' and self-sufficiency had been crucial in making the project a welcome intervention. At the same time, there was an awareness of the need to involve non-Javanese communities in the OC project. The feeling that the project had involved women without fully acknowledging their contributions officially as beneficiaries or participants was also strong.

Analysing Gender with Participants

Women living in villages in and around the mine operation play a number of roles at home, in the communities and as productive agents. At home, they are mothers and housekeepers, cooks and carers, but they also act as producers of a significant amount of food and other crops, tend to fruit trees and livestock. Most women make trade-offs in allocating their time, labour and productive resources between their varied roles and obligations. Most household plots display these mixed responsibilities and combine production cycles where the primary responsibility for the crops can be shared. Almost the entire range of farming activities involves a collaborative effort between women and men. Women are responsible for domestic livestock, vegetables and tree crops around their households. Digging and land preparation are seen primarily as men's tasks, but caring for the soil, weeding and regular watering are women's chores. The economic status of the family has a distinct influence on a woman's involvement in the family farm. Many Javanese women living in Rantau Pulung are conscious of their position in the social hierarchy. Women from more affluent families or women with better educations spend less time in the field and more time cooking and in other tasks within the household. Going out and working in the field is also associated with a darkening of the complexion due to the exposure to the sun and other elements. Social and gender norms tend to be followed more strictly by women who have alternative means that allow them to stay away from the sun.

Some homes in both Rantau Pulung and Sangatta double as small grocery stores or shops supplying provisions to local families. Both the husband and wife may participate in running these small enterprises and gender roles are not always strictly compartmentalised, but may overlap significantly. However, one T&D participant, Ibu S, who sells food at the market to support her family, receives no support from her husband. She was reluctant to comment on the project and was unable to say whether she thought she might possibly make more money through the project. She was generally shy and quiet, with no education, but agreed that she has made some new friends through the program, which has expanded her social network. However, it is important to remember that because Sangatta is urbanised and hosts many migrant families, the boundaries of the households there tend to be less permeable and changeable than those families living in Rantau Pulung.

Since the rural household is the primary site of both production and consumption, we used households as starting points in our focus group discussions. Initial discussion focused on the 'activity locus'— identifying where activities are performed by a family—in the family field, shop or in the outside community. This revealed the extent of women's mobility as compared to men. For community development projects, gender-differentiated mobility has implications for project delivery and reaching the most disadvantaged. Although most projects are assumed to be gender-neutral, they are not targeted to homogeneous populations within a community. As noted by Overholt et al. (1985), the gender-based division of labour, as well as access to and control over resources and benefits are likely to differ within a community. Hence we aimed to develop separate activity profiles for women and men at each of the project sites.

For the United Nations (UN) funded project, Household, Gender and Age, researchers Masini and Stratigos (1991) from the UN University found that many men and women used the 'life-course approach' as the most suitable for capturing the impacts of macro-changes on women and households. They found that talking about changes with individuals within the household was useful even for women with very little education. In our focus groups, we used a version of this method: instead of individual interviews we listened to individual men's and woman's free-flowing narrations. As women recalled their family trajectories they recalled what had happened to them in the course of their lives and began to reconstruct their life histories, and even related major changes in terms of events in the wider environment. The life-course method also revealed the roles, activities and the status of the individual woman or man in relation to the changes that development projects introduced into their lives. Such free-flowing description produces what is described in gender analysis as an 'activity profile' that gives a more or less accurate assessment of the interaction between women and external projects (Overholt et al. 1985). As the UN researchers noted, an important task in the process is 'how one categorises activities conceptually', and that was our main thrust in tracking gendered access to assets.

Three major questions were raised in our gender analysis:

1. Who does what, when and where in the household or home-based enterprise, on or off the farm, and in terms of household maintenance?

2. Who has ownership, control over or access to resources, production, knowledge, technology, time and decision-making power?

3. Who benefits from the existing organization of production, community and household resources?

Discussion of these questions during the focus groups often brought us back to the issue of gender-based roles and responsibilities, and differential control over resources. For example, although women earned independent incomes and managed household farms, they noted that this did not necessarily enhance their value to their husbands. Often, the project targeted men as farmers and receivers of training, leading to an erosion of self-esteem and confidence. For example, one OC participant, Ibu K, felt that the project added to her knowledge and increased her skills, but did not directly address her needs or see her as a recipient of the project's benefits. Ibu K had a baby and when she came to the training sessions or visits by experts, she needed to bring him along. Sometimes she received comments that she should not have brought the baby to public meetings. Although she hardly spoke during the group discussions, she was forthcoming with her views during one-to-one sessions, and described how, in spite of being at the margins of the development project, her life was influenced by it.

Main Agents and Factors of Change

Change at both the individual and the family level was discussed with the participants in the OC workshop. Both male and female participants reported that at the family level, success depends on an effective interaction between the major agents: the farmer; the local government and the Company (KPC).

The diagram we developed, however, pictured the government as a stand-alone entity. This has a historical reference: KPC was able to establish a 'patron–client' relationship with the communities around its operation sites before the Indonesian government arrived at its doorstep through the decentralisation of administration. Figure 7-2 illustrates the continued heavy dependence of the community on KPC's community development programs.

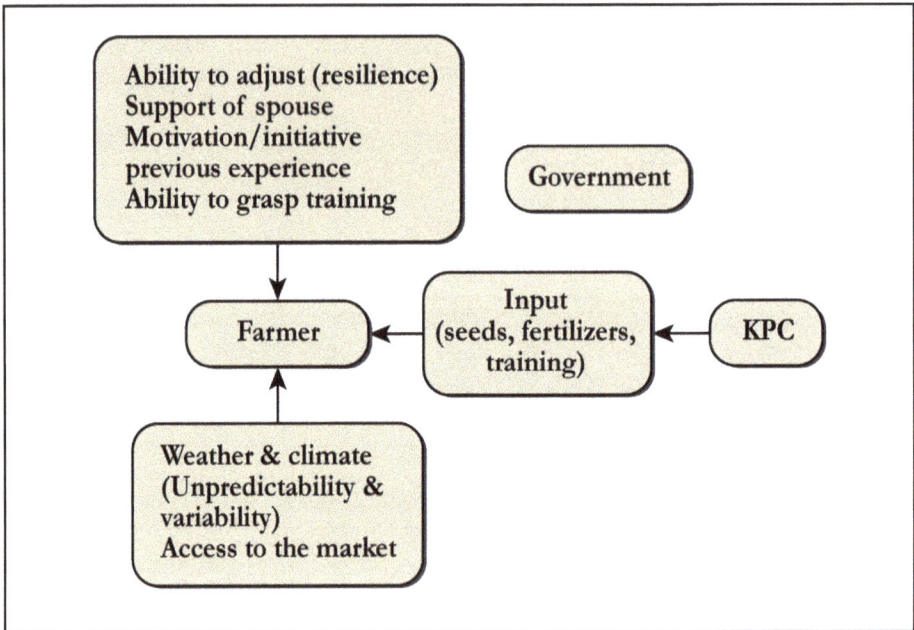

Figure 7-2: Agents and factors of change.

Source: Adapted from various sources.

In the past, innovations in farming techniques grew out of conversations with village elders, who offered suggestions based on their knowledge of traditional farming techniques. With intensive logging, there are now fewer trees available and the water-retentive capacity of the badly eroded soil has declined considerably. As the area has grown drier, the village elders recommended terracing the fields and raising embankments to prevent runoff. At the individual level, participants placed importance on psychological factors like initiative or motivation, and prior exposure, experience or education to take advantage of training. A number of male participants emphasised the support of their spouse as critical for success. Interestingly, many women felt that a 'general support network'—such as the availability of neighbours or relatives for child care— was an important factor besides prior training and knowledge or exposure.

Success Factors in Asset-Building

To follow up, we discussed factors for success. The participatory exercises generated a significant amount of discussion between women and men. These exercises drew heavily on the creativity of individual participants, with us using stories and examples to break the silence on issues relating to women's and men's work burdens. 'What are the factors for success?' was a theme that was debated extensively, and there were significant differences between women and men in accounting for success. Saifuddin, one of the more successful farmers, observed wisely: 'Success is relative, the most important thing is that I have learnt something new. The land that was lying fallow around their houses has now been put to use for the families, and earning extra incomes'. For many other men like him, success meant being able to 'put to use' land that was 'lying fallow'. Women, on the other hand put more importance on 'learning something new', and pointed out that learning a new skill gives them the opportunity to use it to add to the family's income. Ibu H, for example, is a middle-aged single woman whose husband died at an early age, leaving her with the responsibility of bringing up three children. She took up a tailoring course and sought money from a local moneylender to start her own tailoring business. Now she makes employee uniforms and embroiders the company logos for KPC as well as its local contractors.

Word Pictures: 'Before–After' Asset Differences

At the next stage, we developed word pictures based on the worldviews of the participants. The outcomes were practical and realistic and helped us to take up a rapid appraisal to compare the statements with reality, allowing the community to own the method. The 'worst picture' of a household's livelihood status was envisaged as little or no land, a limited amount of food available from the farm, poor-quality land, droughts destroying the orange plants, slash and burn causing forest fire, the presence of illegal logging and having no livestock. The 'best picture' of a household was envisaged as more land, more produce from land, good orange crops, greater access to markets, own transport and good rainfall. Following the discussions in the workshops, the participants themselves drew a 'before' and 'after' picture in terms of their own livelihood status (see Table 7-1).

Table 7-1: Before–after word pictures.

Before-After (Impact)	Before	After
Best off	–	Saifuddin, Tris
Better off	Tris	Basuki, Bai, Wahidin, Imam, Yohanis, Sugiran
Medium (no change)	–	Remi, Said, Mukhroji, Vincent
Worse off	–	–
Worst off	Basuki, Dul, Mukhroji, Vincent, Yohanis, Sugiran	Dul

Source: Author's own data.

At this stage in the LAST method, we prepared an assessment sheet to compare it with the reality of a particular household. This involved a rapid assessment process to find which description or collection of indicators fitted closest to the reality of the individual woman and man within a household. For this process, individual questions were not asked for every indicator; rather we used our judgment on the relevant part and combined direct observation with semi-structured questioning. In the field, this took approximately 30 minutes or so for a single household for the LAST enumeration. The LAST sheet was then used as a checklist and for recording the closest situation to the household and did not intrude on the interview.

Changing Self

Overall, the women in villages who had participated in company-sponsored projects felt that they flourished as individuals; one noting she was 'like a changed person'. One participant, Ibu S, observed that before she started the business, she could not imagine that it was possible for her to undertake such a venture. But now she thinks: 'If I could do this business—rather than staying at home—I could earn more.' The comments echoed the sentiments of feeling 'richer because of more knowledge' and feeling 'increased enthusiasm'. This was true in particular of participants in the OC group from which one person had been sent for intensive training in Bandung. One woman made the observation that, 'now that I have learnt something, I want to do it better.' A number of women expressed a greater understanding and appreciation of self-reliance through the interviews. One said, 'So far KPC has provided the materials but I want to buy them myself.' Another said, 'I want to work continuously to make sure that my children are better off.'

More so than men, women participants criticised the short life spans of individual development projects, which tend to be funded in small bursts as funds are made available on a project-life basis by the company. Another point made often by women was the need for more capital to be made available directly to the individual ('will I get credit to run the business?'). These needs reflect Indonesian social institutions that make women's contributions invisible— women did not own land, lacked education and were neglected in training programs, and hence did not receive credit.

The manner in which women have dealt with these systemic shortcomings is innovative and clearly reveals the value of an asset-based approach. Traditional 'women's groupings' such as Qur'an study groups or book-reading groups provide women with the necessary tools to deal with market-based processes that attempt to leave them by the wayside. A significant role is played in rural communities by a traditional grouping of households into desa wisma, or families of ten. They act as vital support networks in swiftly changing rural contexts like the one we were working in—experiencing rapid urbanisation; an influx of immigrants and changing cultural and natural landscapes (changes that are very common in remote areas that experience rapid changes because of mining-led development). Women in these groups of ten families help each other in various ways—not only by lending crucial capital resources during a wedding or medical emergency, and to set up businesses, but also by helping each other by looking after each other's children and caring for each other's families.

Important lessons were learnt in the course of the research by all participants. True to the essence of action-research, our research certainly had significant impacts on those who took an active and engaged part in it. Not only did the male bias in the company's development projects become apparent to all concerned, a number of community extension personnel acknowledged the new and gendered way in which they began to see what, to them, was previously a homogeneous and externalised 'community'. This bias was neither explicit nor intentional, but it operated nevertheless in favour of men as a gender, and against women as a gender. A number of men contributed substantially to the research process and once identified, were interested in combating the bias. The process of research was also transformative for us, participants from outside. It allowed us to understand the important roles played by women and the contributions they made in building the household and making the community. It also showed that alternative ways of seeing the world are possible; that poor and rural women have a number of valuable support systems that need more nuanced understanding and enhancement. This is a vital step in making development projects beneficial to both women and men. Once recognised, the asymmetries in the lived experiences of women and men, seen in the course of research through the LAST method, can be addressed and combated in time. Further, the complementarities in women and men's roles be made equitable and just, and the strengths that women depend on can be supported.

References

Adato, M. and R. Meinzen-Dick, 2002. 'Assessing the Impact of Agricultural Research on Poverty using Sustainable Livelihoods Framework.' Sussex: IDEAS.

Ashley, C. and K. Hussein, 2000. 'Developing Methodologies for Livelihood Impact Assessment: Experience of the African Wildlife Foundation in East Africa.' London: Overseas Development Institute (Working Paper 129). Viewed 17 July 2010 at http://www.odi.org.uk/resources/download/2032.pdf

Barry, M.M. A. Doherty, A. Hope, J. Sixsmith and C. Cecily Kelleher, (2000). 'A community needs assessment for rural mental health promotion.' *Health Education Research* 15(3): 293–304.

Bond, R. and N. Mukherjee, 2002. 'Livelihood Asset Status Tracking: An Impact Monitoring Tool?' Journal of International Development 14: 805–15.

Chambers, R., 1989. 'Vulnerability: How Do Poor People Cope?' *IDS Bulletin* 20(2): 1–8.

Elson, D. (ed.), 1991. *Male Bias in the Development Process*. Manchester and New York: Manchester University Press.

FAO (Food and Agriculture Organisation) and ILO (International Labour Organisation), 2009. *The Livelihood Assessment Toolkit: Analysing and Responding to the Impact of Disasters on the Livelihoods of People, Rome and Geneva*. Rome and Geneva: FAO and ILO.

Goulet, L., 2008. 'The Road to Self-Reliance—ABCD, Case Study.' In D. Green (ed.), *From Poverty to Power: How Active Citizens and Effective States Can Change the World*. London: Oxfam International.

Guy, S.M., (1996). *Community needs assessment survey guide*. Utah: Utah State University Extension.

Kabeer, N. 2002. 'From Feminist Insights to an Analytical Framework: An Institutional Perspective on Gender Inequality.' In N. Kabeer and R. Subrahmanian (eds), *Institutions, Relations and Outcomes: A Framework and Case Studies for Gender-Aware Planning*. New Delhi: Kali for Women.

Koch, T., 1994. 'Beyond Measurement: Fourth Generation Evaluation in Nursing.' *Journal of Advanced Nursing* 20: 1148–155.

Lahiri-Dutt, K., 2004. 'Gender Needs Assessment Report.' Canberra: ANU. Viewed 20 January 2011 at http://empoweringcommunities.anu.edu.au/documents/LahiriDutt_2004_GenderNeedsAssessment.pdf

Massini, E. and S. Stratigos (eds), 1991. *Women, Households and Change*. Tokyo: United Nations University Press.

Nicole, A., 2001. 'Adopting a Sustainable Livelihoods Approach to Water Projects: Implications for Policy and Practice.' Overseas Development Institute, London (ODI Working Paper 133).

Overholt, C., M.B. Anderson, K. Cloud and J.E. Austin (eds), 1985. *Gender Roles in Development Projects: A Case Book*. West Hartford: Kumarian Press.

Turton, C., 2001. *Livelihood Monitoring and Evaluation: Improving the Impact and Relevance of Development Interventions*. Sussex: Institute of Development Studies.

UNDP (United Nations Development Programme), 2002. *Handbook on Monitoring and Evaluation for Results*. New York: UNDP Evaluation Office.

Whitehead, A., 2002. 'Tracking Livelihood Change: Theoretical, Methodological and Empirical Perspectives from North-west Ghana.' *Journal of Southern African Studies* 28(3): 575–98.

8. Women-Owned SMEs in Supply Chains of the Corporate Resources Sector

Ana Maria Esteves

Introduction

The creation of economic opportunities through the development of small and medium enterprises (SMEs), and integration of these enterprises into the supply chains of large companies, has been promoted by international development agencies such as the International Finance Corporation (IFC) and the United Nations Industrial Development Organization (UNIDO). A strong SME sector is considered to provide social benefits such as the empowerment of local communities and a path out of poverty. In developing countries, SMEs are also considered to have a positive impact on regional income distribution. Their labour intensiveness with low technological and capital equipment requirements provides employment for relatively disadvantaged members of the community, such as women with reproductive and household duties, the unskilled and the elderly (UNIDO 2003, Nelson 2007).

Reinforcing the message of the development agencies is a growing recognition within mining, oil and gas companies of the uneven revenue distribution and limited employment opportunities offered by the sector (Esteves and Vanclay 2009). This has led to some consideration of the social and economic value potentially realised though the sourcing, sub-contracting and procurement of goods and services from local small enterprises. The management complexities of what is commonly referred to as 'local content' are numerous. Companies increasingly face contractual obligations under host country agreements for local content and social investment. However, local SMEs are often excluded from procurement processes, due to the large scale of contracts, inaccessible procurement systems and complex technology, safety, labour and reporting requirements. Where local content is pursued, activities are often constrained by infrequent management attention, market failures and governance gaps (ODI 2005; Wise and Shtylla 2007).

Adding to the disadvantage encountered by SMEs is the perception among financial institutions or potential investors that SMEs are high risk borrowers or low profit investments, and are typically characterised as having insufficient assets and low capitalisation, and therefore vulnerable to market variability. The shortcomings are weighted with associations with high mortality rates, lack of accounting records and business plans and high transaction costs (UNCTAD 2004, Ruffing 2006).

Women entrepreneurs face additional restrictions: cultural values and social norms often limit equal participation in society; they have unequal access to productive resources and services, such as finance and skill upgrading; certain legal provisions and legislative systems prevent women from securing property rights (for instance the husband will have legal ownership of the business, even if the person responsible for running the business is a female), and they have reduced opportunities as a consequence of their reproductive role (De Groot 2001; UNIDO 2003). Literacy levels can be a major challenge in developing countries, where some traditional societies constrain the inclusion of females in entrepreneurship development. Mature women are more likely to have been excluded from formal education, reducing their chances of becoming formal entrepreneurs.

This chapter considers these issues by taking the perspective of corporate professionals seeking to address the following management challenge: how can supply chain value be enhanced through building the capacity of female entrepreneurs to participate in SME development activities, while also enhancing development benefits to local communities? The chapter commences with an overview of concepts related to 'local female content' as practiced by mining, oil and gas companies. The concepts are illustrated through an example involving a resources company operating in Ghana that has partnered with the IFC and a local business association to develop business linkage and mentoring programs. Finally, knowledge gaps in the supply chain management literature and corporate practice are highlighted.

Integrating Concepts of Supply Chain Management and Women's Development

The 'supply chain' is a relatively new management term and concept, having developed mainly as an extension of production and operations management over the past 20 years. Whereas production and operations management concentrated on the firm's efficiency, quality, service, delivery performance and other aspects of managing assets for performance, supply chain management recognises the importance of coordination and control across firms that are participating together in supplying a set of goods or services. Supply chain management considers that it is not only firms that produce and compete, but that 'chains' of companies that supply each other can be considered together, as a group of coordinated owners of assets locked into a mutually beneficial relationship. It refers to the flows of goods and services, information and money between inter-related partners in production of these outcomes.

The supply chain management literature points to numerous variables which influence the relationship between the company and prospective local suppliers. Those summarised by Croom et al. (2000) include:

- the sourcing strategy, for example, sole-sourcing, multi-sourcing, or partnering;
- the attitude and commitment to collaborative improvement programs;
- the positioning of the local firm within the total network;
- the extent of dependence on the network, measured as the proportion of a supplier's business which is dedicated to the supply network;
- the longevity of the relationships, for example, past behaviour, opportunism and trust in suppliers;
- the technological or process links, for example, if the supplier holds the equipment needed to make their customer's product, or the existence of electronic links to facilitate ordering and payment;
- the existence of legal ties, such as contracts or shared patents, for example;
- the degree of power and influence of each party,
- the length and complexity of the chain; and
- the distance from the end-customer (the greater the number of stages, the less an organisation will perceive its dependency upon end-user demand).

Supply chain and procurement managers typically promote and facilitate linkages with local firms by adopting a range of approaches. These have been discussed in some detail by Engineers Against Poverty (2007) and are as follows. The most common means of implementing local content policies is 'preferencing' specific companies or groups in the award of tenders. Typically, an enterprise will have the preference figure discounted from its tender price. South Africa has been one example, where Black Economic Empowerment (BEE) legislation was introduced to address inequalities created by past history. Under the legislation, many companies are obliged to procure preferentially from previously disadvantaged black-owned enterprises.

Another approach is 'set aside' or 'reservation'. Local SMEs can be disadvantaged or even excluded from supply chains as a result of the increasing industry trend to 'bundle' supply requirements into larger (and therefore fewer) contracts, in order to reduce transaction costs and enable suppliers to benefit from bulk discounts. Unbundling can allow goods and services deemed suitable for local supply to be set aside or reserved for local enterprises.

Some companies prefer to incentivise and transfer many of their responsibilities for local content to a better-prepared lead contractor. 'Pushing' responsibility for local content down the supply chain can occur through the use of prequalification processes to ensure that major contractors have the required capabilities in capacity-building, subcontracting or partnering with local firms. At the tendering stage, companies can specify that potential contractors enter into a joint venture with a local enterprise, sub-contract portions of the contract to a local enterprise or source from local enterprises. Through tendering processes, companies can even preference large contractors that develop innovative approaches to maximising local enterprise involvement.

A further strategy involves adapting procurement processes so as to ensure their accessibility to local SMEs. This can be achieved by communicating future demand for goods and services to allow suppliers enough time to build the required capability and having longer contract periods to enable the acquisition of capital equipment. Tender processes can also be made more accessible by simplifying pre-qualification processes and tendering procedures, having appropriately translated and distributed tender documentation and offering training and assistance in the tender preparation process and to understand the company's standards and requirements. Supply chain managers and technical end users can also provide support to supplier development programs typically delivered by external organisations or dedicated internal units. Assistance is provided in focusing program activities through mapping future demand and establishing the criteria for assessing goods and services.

To ensure that women benefit from enterprise development opportunities arising from company operations, management approaches need to be examined from a gender-aware development perspective. Potentially, major resources companies provide a significant opportunity to stimulate the participation of women in local economies—by integrating SMEs with female ownership and employment into the corporate supply chain, and building the capacity of these SMEs to become competitive. However, economic growth does not affect men and women equally. Genderless approaches to SME development will ignore the complexities that underlie exchanges between women and men (Srinivasan and Mehta 2003). Any industry intervention that aims to bring the benefits of economic growth to women will require an understanding of women's vulnerability, their socio-cultural context, the institutional structures and processes within their communities and the strategies employed by women to manage their assets.

More consideration of 'what works' is warranted in this little researched area of practice. There has been limited empirical evidence of the promoted benefits of supply chain linkages to women's development. Typically, these linkage programs address areas such as financing, business management and technical skills development, and support and institutional strengthening activities to encourage an enabling environment and reduce gender gaps in legal, business and financial systems (De Groot 2001; ODI 2005; Wise and Shtylla 2007). The potential benefits of SME linkages for companies, SMEs and the local region have been described by Jenkins et al. (2007), Ruffing (2006), UNIDO et al. (2004) and UNCTAD (2001). However, not all linkages between major corporations and local SMEs have been shown to create beneficial development impacts. An UNCTAD (2001) report pointed out that the extent and nature of the impact depends on the amount of procurement a company sources from the region, the size of the company's activity relative to the regional economy and the socio-cultural, regulatory and political context. In highly protected economies, there is little incentive to upgrade suppliers' technology. This fosters supplier dependence, placing it at risk if the company goes through a downturn, changes procurement strategy or decides to decommission.

From a corporate perspective, while in theory SME linkages are beneficial by offering savings in delivery and price, lower labour costs associated with poor regions and the close proximity of local suppliers to a project; in reality, there are a number of issues. Attaining the necessary quality standards can be expensive and technical, business management, financial support and risk guarantees may be necessary (ODI 2005). In imperfect market conditions, the preferencing approach can lead to local enterprises charging monopolistic prices. With regard to the set aside approach, while it facilitates entry for SMEs, there are also disadvantages for companies, such as additional administrative costs, potentially reduced effectiveness of project management and increased contractual risk. A further issue affecting all local content approaches is the problem of token (also referred to as 'front') organisations being established in order to win business (Engineers Against Poverty 2007).

Companies also face internal pressures which discourage local content. An Overseas Development Institute (ODI) report (2005) showed how the industry trend towards global strategic sourcing has led to favouring low cost/high quality suppliers, which are often located in areas other than local communities. The report argued that local linkages are enabled by decentralised decision-making and local management with delegated authority for independent sourcing. Further, it is also often the case that enterprise development opportunities are managed by lead contractors rather than the company. Short-term, fixed price contracts can be barriers for lead contractors against participating in meaningful linkages activities.

Ahafo Linkages Program: Promoting Female Participation in Supply Chains

There is general acceptance that developing competitive small enterprises requires coordinated intervention among the private sector, public sector, non-government organisations (NGOs) and public policy. As a result, a number of partnerships have emerged between governments, support institutions and development agencies to establish linkage programs. One such example of SME linkages in the resources industry is the Ahafo Linkages Program (ALP), Ghana; a partnership between Newmont Ghana Gold Ltd (NGGL) and the IFC. The brief overview and analysis of emerging issues provided in this section have been prepared drawing on NGGL documents and interviewing ALP staff.

NGGL Ahafo is a greenfield mine that commenced operation in July 2006. Located in the Brong-Ahafo region of Ghana in West Africa, the mine was the first large-scale economic activity to be undertaken in the region. The region covers an area of just under 40 000 square kilometres and accommodates approximately 1.8 million people who are historically reliant on subsistence farming and small-scale commercial farming. The area is known for the production of cocoa, timber and cashew (Newmont Ghana Gold Ltd 2009).

The development of the Ahafo mine has brought profound social and economic changes to local communities, such as access to new employment and economic opportunities, impacts to existing patterns of settlement and movement and increased demands on community representatives and social institutions, such as local government and civil society organisations (Kapelus and Rogers 2008). In a context of poverty, local expectations over the flow-on economic benefits from mine development are high. The presently low community capacity to service the needs of the mine creates tensions which the ALP in part has been designed to address. The ALP's goals are to maximise procurement of goods and services from seven affected communities, diversify local economic activity and provide management and technical skills training within and beyond the supply chain. Emphasis is placed on strengthening the Ahafo Local Business Association (ALBA). NGGL has a dedicated unit in the supply chain department to work with local suppliers. The internal unit is responsible for maintaining a database of pre-approved bidders for Newmont contracts, establishing local procurement procedures and obtaining internal 'buy-in' within the company for local procurement. The ALP assists in bringing pre-approved bidders up to Newmont standards.

While the ALP is still in the early stage of development, there is evidence pointing to increased economic benefits associated with the supply of goods and services to the mine to both men and women. Table 8-1 has drawn on a summary report prepared by ALP Program Management Office in January 2009, which included source data prepared by the NGGL Local Supply and Community Development unit). As shown in the table, the number of local SME suppliers in 2008 totalled 125. Participation of female-owned companies in Ahafo's local content amounted to 17 per cent, or 21 companies. This participation level increased from four companies in 2006.

Table 8-1: NGGL Ahafo Local Content Indicators 2006–2008.

Indicator	Baseline 2006	Value 2007	Value 2008
Contracts awarded to local vendors (USD)	1 718 949	4 182 654	4 668 404
Contracts awarded to female local vendors (USD)	302 003	707 671	950 537
Proportion of contracts awarded to female local vendors (%)	18	17	20
Number of local vendors awarded with contracts	25	52	125
Number of female local vendors awarded with contracts	4	13	21
Proportion of female local vendors awarded with contracts (%)	15	25	17

Source: Author's own data.

The female participation indicators reported in the table show the natural evolution in local female content over a period of three years, 2006 to 2008. While the share of contracts awarded by NGGL to female-owned vendors has not varied significantly over this period (ranging from 17 per cent to 20 per cent), the number of female vendors has grown from four to 21 and the direct economic benefit to local female vendors has more than tripled. In 2008 the ALP decided to try and connect more women with ALP activities, thus ensuring equal access to business opportunities for both men and women. This reflected an attempt to increase total levels of local content, rather than increasing the share of available contracts to female vendors. Both male and female entrepreneurs were still required to go through the same screening and selection processes.

An assessment was conducted by ALP in early 2008 of women's economic activities and female entrepreneurs in the local area. As a result of these efforts, the company was able to access new local sources of goods and services. By December 2008, five female-owned companies had completed training with the ALP. One female-owned company (out of the 22) completed the Local Supplier Development training, which involved technical assistance to and mentoring of 18 females (representing 44 per cent of individuals). Local Economic Development training involved four female-owned companies (out of the 25) and training of 14 women (representing 56 per cent of individuals undergoing mentoring and training). By the end of 2008, ALP also participated in Institutional Capacity Building activities which involve Training of Trainers (two women out of the six local consultants hired to provide short-term training to ALBA members). In the same year, short-term training sessions were conducted for ALBA members in Financial Planning (six women among 35 participants), Basic Costing and Pricing (nine women among 29 participants) and Tender Management (six women among 30 participants).

At the time of writing this chapter, two out of three women had dropped out of the Local Supplier Development training program. No single dominant reason was identified by program staff: the lack of completion could have been attributed to the lack of basic levels of education, low levels of interest in running a business (an activity attributed lesser priority in the local traditional society) and the quality of mentoring. It was also noted by staff that the absence of strong formal entrepreneurship activity and prevalence of weak business practices are barriers which are not restricted to the women in the area. ALP had made some adaptations to the mentoring program in response, and more tangible results were being sought for the second group of program participants. A conclusion that could be drawn from this experience is that the absence of 'exit surveys' or another suitable evaluation mechanism means a missed opportunity to start tracking the program in terms of its broader social effects at the individual, family and community levels, and particularly its effects on gender equity.

A further challenge facing the ALP is the dependency of local suppliers on NGGL contracts, making them vulnerable to market fluctuations. The ALP is attempting to address this issue by providing assistance in diversifying the markets of companies being mentored. A new initiative, the Ahafo Business Week, involves a business fair promoting suppliers and entrepreneurial activity associated with the establishment of the NGGL Ahafo mine to other companies operating in the area. The first fair conducted in April 2009 involved 30 local businesses showcasing their products and services.

ALP staff also identified the need for more dissemination of ALP experiences. Capturing and communicating the learning is of particular significance when one considers the 2010 timeframe for IFC's exit from the ALP partnership. While a succession plan for a program manager is in place, NGGL has yet to make a decision as to the means by which the linkages program will be replicated. Two models being considered involve situating the program either within the company's internal community development unit, or at arms' length from the company under the Ahafo Sustainable Development Foundation. Both options present trade-offs as to program sustainability, community capacity-building orientation, and opportunities for corporate-community engagement. One of the drawbacks of not having a clear exit strategy is that the learning curve for those who will take over the management of the program will be steeper when the IFC exits than would have been the case if the exit strategy had been in place from the commencement of the ALP.

Conclusion

Corporate procurement and supply chain management provide the opportunity to address 'the female face of poverty' through sustainable enterprise development. However, there is a number of implementation issues faced by corporations. These are well known; for instance, Jenkins et al. (2007) have summarised the challenges in establishing linkages as: securing internal commitment; obtaining information about SMEs in the local market; measuring the impact (particularly societal impact) of linkages; managing community expectations; reducing the economic dependence of local SMEs on the company; building the skills and capacity of SMEs and improving access to finance. The management discipline would benefit by further research into strategies that assist managers in addressing these issues, selecting appropriate linkage strategies, and building an understanding of the enabling environment for sustainable local enterprise development.

Measuring the impact of linkages is a prominent gap within the resources sector. Here, the purchase of goods and services tends to be assessed in terms of their economic contributions to regions. These activities are rarely assessed in terms of their direct and indirect social impacts and potential development opportunities. A broader view of impacts is needed to build the empirical basis for informing management action (Esteves 2008), and to increase the capacity of managers in making decisions that enable local women to bid for and work within, and beyond, the resources sector. This is reinforced by increasing stakeholder expectations that companies demonstrate their impacts to the regions in which they operate.

Enhancing the development benefits to women will also require a shift from perceiving support for female-owned SMEs merely as a means of increasing the overall opportunities to entrepreneurs in affected regions (as is evident in the early stages of the ALP), to also being a means of addressing inequalities in households and communities. With this shift, gender-sensitive evaluations of SME program performance will become important; with indicators that are sensitive to local understandings of women's access to and control over resources, and that go beyond those explicitly stated by customary law. Such indicators are best developed through participatory approaches to program evaluation, where female entrepreneurs are involved in determining and evaluating the positive and negative effects on themselves, their households and livelihoods.

References

Croom, S., P. Romano and M. Giannakis, 2000. 'Supply Chain Management: An Analytical Framework for Critical Literature Review.' *European Journal of Purchasing and Supply Management* 6: 67–83.

De Groot, T.U., 2001. 'Women Entrepreneurship Development in Selected African Countries.' Vienna: United Nations Industrial Development Organization (PSD Technical Working Paper 7). Viewed 3 January 2009 at http://www.unido.org/index.php?id=o4810

Engineers Against Poverty, 2007. 'Maximising the Contributions of Local Enterprises to the Supply Chain of Oil, Gas and Mining Projects in Low Income Countries: A Briefing Note for Supply Chain Managers and Technical End Users.' London: Engineers Against Poverty. Viewed 3 January 2009 at http://www.engineersagainstpoverty.org/news.cfm?cit_id=3989

Esteves, A.M., 2008. 'Mining and Social Development: Refocusing Community Investment Using Multi-Criteria Decision Analysis.' Resources Policy 33(1): 39–47.

Esteves, A.M. and F. Vanclay, 2009. 'Social Development Needs Analysis: Adapting SIA Methods to Guide Corporate-Community Investment in the Minerals Industry.' Environmental Impact Assessment Review 29(2): 137–45.

Jenkins, B., A. Ahalkatsi, B. Roberts and A. Gardiner, 2007. 'Business Linkages: Lessons, Opportunities and Challenges.' Cambridge: IFC, International Business Leaders Forum and the Kennedy School of Government, Harvard University. Viewed 3 January 2009 at http://www.hks.harvard.edu/m-rcbg/CSRI/publicątions/report_16_BUSINESS%20LINKAGESFINAL.pdf

Kapelus, P. and G. Rogers, 2008. 'Global Community Relations Reviewed: Site-based Assessment of Ahafo Mine, Ghana.' Oxford: Synergy Global Consulting. Viewed 26 April 2009 at http://newmontghana.com/index.php?option=com_content&task=view&id=224&Itemid=2

Nelson, J., 2007. 'Building Linkages for Competitive and Responsible Entrepreneurship.' Cambridge: UNIDO, Harvard University. Viewed 3 January 2009 at http://www.unido.org/fileadmin/user_media/ Services/PSD/CSR/Building_Linkages_for_Competitive_and_Responsible_Entrepreneurship.pdf

Newmont Ghana Gold Ltd, 2009. 'Ahafo Mine.' Viewed 3 January 2009 at http://newmontghana.com/index.php?option=com_content&task=view&id=12&Itemid=35

ODI (Overseas Development Institute), 2005. 'Levers and Pulleys: Extractive Industries and Local Economic Development—Incentivising Innovation by Lead Contractors Through Contract Tendering.' London: ODI (Briefing Note 3). Viewed 3 January 2009 at http://www.odi.org.uk/resources/specialist/ business-development-perfomance-briefings/3-extractive-industries-local- economic-development.pdf

Ruffing, L., 2006. Deepening Development through Business Linkages. New York: United Nations.

Srinivasan, B. and L. Mehta, 2003. In H.A. Becker and F. Vanclay (eds), The International Handbook of Social Impact Assessment: Conceptual and Methodological Advances. Cheltenham: Edward Elgar.

UNCTAD (United Nations Conference on Trade and Development), 2001. *World Investment Report 2001: Promoting Linkages*. New York: United Nations.

————, 2004. 'Business Linkages: Roster of Good Practices.' Geneva: UNCTAD.

UNIDO (United Nations Industrial Development Organization), 2003. *A Path out of Poverty: Developing Rural and Women Entrepreneurship*. New York: UNIDO.

UNIDO, Deloitte, The Global Compact and UNDP, 2004. 'Partnerships for Small Enterprise Development.' London and Washington: Deloitte Touche Tohmatsu Emerging Markets Ltd. Viewed 3 January 2009 at http://www. unido.org/index.php?id=o44527

Wise, H. and S. Shtylla, 2007. 'The Role of the Extractive Sector in Expanding Business Opportunity.' Cambridge: Harvard University (M.A. thesis). Viewed 3 January 2009 at http://www.hks.harvard.edu/m-rcbg/CSRI/publications/ report_18_EO%20Extractives%20Final.pdf

9. On the Radar? Gendered Considerations in Australia-Based Mining Companies' Sustainability Reporting, 2004–2007

Sara Bice

Introduction

Australia-based mining companies[1] are inarguably devoting more attention and funds to sustainable development—certain companies now contribute one per cent of pre-tax profits to community programs and most undertake some form of sustainable development programs. The social and environmental impacts of mining, however, remain significant and are frequently negative (Atkinson and Community Aid Abroad Australia 1998; Auty 1993; Smith 2008). Academics, mining company representatives, governments and non-government organisations have devoted great attention to recording and analysing mining's impacts (For just a few of many examples, see: Emberson-Bain 1994; Klubock 1998; Vittori et al. 2006; Jones et al. 2007; Walton and Barnett, 2008). The book's focus on 'Mining, Gender and Sustainable Livelihoods' directs our attention to this range of issues—from land ownership and management to changing family or community structures, to increased prostitution or the spread of HIV/AIDS—mining has significant and abiding gendered impacts (Storey 2001; Kunanayagam 2003; Vittori et al. 2006).[2]

Often, when we consider actions taken to address mining's gendered impacts, we think of the work of non-government organisations, local governments or intergovernmental organisations, such as the United Nations Development Fund for Women (UNIFEM). This chapter takes a different perspective by looking at how mining companies themselves understand and address gender issues.

---

1   Australia-based mining companies are defined as those companies which are either Australian Stock Exchange listed or which have regional headquarters within Australia, combined with significant Australian and Pacific operations. All companies studied in this chapter are referred to as 'major players' in Australia by IbisWorld (2007).

2   I am grateful to have been given the opportunity to discuss an earlier version of this work at the ANU/World Bank *Mining, Gender and Sustainable Livelihoods* workshop, 6–7 November 2008. Special thanks to Dr Kuntala Lahiri-Dutt and Ms Sophie Dowling for their organisation and leadership of the workshop. Thanks also go to my Ph.D. supervisor, Dr Tim Marjoribanks, University of Melbourne, for his helpful comments on various versions of this chapter.

It examines the extent to which Australia-based mining companies report on gendered issues and how those issues are addressed under the auspices of 'sustainable development' or 'corporate social responsibility' (CSR) programs. To do so, the chapter analyses four years of sustainability reports from five leading Australia-based mining companies to create a snapshot of these companies' contemporary conceptions about and actions on gender issues. The content of sustainability reports is further explored through the application of auditing literature, which helps us to theorise why mining companies address the issues they do, in the ways that they do, through sustainability reporting.

Three central questions shape the study. Firstly, what are Australia-based mining companies putting into the public domain about gender issues, relative to their operations? Secondly, what does this information tell us about the ways in which companies are approaching, prioritising and engaging with gender issues? Thirdly, why might companies address gender issues in certain ways?

The study employs content and discourse analysis techniques to provide a baseline measure of the extent to which gendered considerations are 'on the radar' of Australia-based mining companies. The content analysis provides insight into the degree to which these companies are explicitly incorporating gender issues into their sustainability programs. The discourse analysis further reveals the ways in which gender is understood, approached and situated within sustainable development programs. This analysis also identifies company interests and priority concerns, including key shareholder and stakeholder concerns.

This chapter asserts that sustainability reports reveal a cyclical relationship between internal reporting indicators, voluntary reporting initiatives, mining companies' sustainability programs and sustainability reporting. Specifically, it is suggested that the paucity of gender indicators within commonly used voluntary reporting initiatives, such as the Global Reporting Initiative (GRI), contributes to a cycle wherein gender does not make it onto sustainability program agendas, partly because it is an issue that does not appear to require reporting. In other words, limited gendered reporting indicators arguably influence the prioritisation of gender issues on mining companies' sustainable development agendas, which contributes to a lack of gender-focused sustainability programs, which normalises the marginalisation of gender issues; and so the cycle continues.[3]

---

3   Although reporting indicators appear first in the way the cycle is described here, it is difficult, if not impossible, to determine where the cycle begins. What is clear, however, is that a relationship between these factors exists yet remains relatively unstudied.

Growth of Sustainability Reporting in Mining

Increasingly, Australian mining companies seek to address the social, environmental and economic impacts of their operations on communities through 'corporate social responsibility' or 'sustainable development'[4] programs and policies. Sustainable development activities range from adoption of quasi-governmental roles (for example the provision of water sanitation infrastructure) to employee giving programs. These programs frequently target areas of perceived community need (such as HIV/AIDS education) or attempt to address socio-cultural issues (such as recognition of sacred sites on mining lease land). Since the early 2000s, many leading mining companies worldwide have produced stand-alone sustainability reports,[5] usually released alongside traditional annual reports. These reports detail companies' sustainable development work and cover a wide range of environmental, occupational health and safety, community health and social issues. In many instances, the reports are used by companies as a vehicle for 'making the business case' for sustainable development to company shareholders and other key stakeholders (Wheeler and Elkington 2001).

Sustainability reports are also used as a kind of 'offensive' public relations tool, wherein companies proactively take the opportunity to disseminate information about the work they are undertaking to mitigate negative environmental and social impacts of mining operations.[6] Before sustainability reports were commonplace, it was especially unusual for companies to report on their community activities at operation sites and relatively unusual for them to report on their environmental activities. It was more likely that information would be released reactively, principally when there was an environmental or social crisis requiring response (Wheeler and Elkington 2001). With the advent of sustainability reporting, mining companies (and major corporations in many other industries) created a medium through which to provide regular disclosure of both positive and negative effects of operations on communities

---

4   The term 'corporate social responsibility' was first used in the early 1950s and remains highly contested. Mining companies tend to use the terms 'corporate responsibility', 'sustainable development', and 'sustainability' to refer to those policies, programs and activities which aim to mitigate negative social and environmental impacts of their operations, or to make a positive contribution to the communities in which they work. For purposes of this chapter, and in alignment with the majority of reports examined here, the term 'sustainable development' will be used.

5   These reports come under a range of names, including: 'Health, Safety, Environment and Community (HSEC) Reports', 'Sustainability Reports' and 'Corporate Responsibility Reports'. Regardless of title, these reports hold in common a focus on environmental, occupational health and safety, and community issues (as opposed to the more financially focused nature of 'annual reports'). The term 'sustainability reports' will be used throughout the chapter to refer to any reports falling under this definition.

6   It is beyond the scope of this chapter to enter into the debate over whether sustainability reports (and related sustainability programs) are mere public relations exercises, however, it is important to note that there still remains much heated debate over the sincerity of company-run sustainable development programs throughout many industries, not just mining (see for example Frynas 2005: 176).

and the environment. For the purposes of this study, the reports also provide an important evidence-base for examining the social and environmental concerns companies prioritise and, specifically, the way they understand and attend to gender issues.

What Does Reporting on Sustainable Development Look Like?

Readers unfamiliar with the field of CSR may wonder why sustainability reports are such a big deal, and, more importantly, what do these reports actually look like? For the most part, sustainability reports are written by corporate communications departments and present a carefully designed, professionally produced image, similar in appearance to the more common annual financial reports. Although there is no formalised or legislated approach to sustainability reporting, reports across industries exhibit key similarities. These reporting conventions have developed in large part because sustainability reporting has become an increasingly competitive aspect of corporate relations (KPMG International and SustainAbility 2008). Sustainability reports are being used to communicate about corporations' competitive advantages, leading-edge practices and risk management in non-financial areas (SustainAbility and UNEP 2006). The use of popular voluntary initiatives, such as the GRI, to guide reporting has also shaped reporting conventions.

Sustainability reports are publicly available documents and are increasingly accessible online (Wheeler and Elkington 2001). The reports are generally structured around the three components of the 'triple bottom line': economic, environmental and social concerns (Elkington 1997). Contents are also heavily based on whichever indicators a company deems relevant[7] and usually present a wide range of charts and graphs detailing the company's performance against internally and externally set voluntary performance indicators. Much of the information provided is quantified, and text is accompanied by full colour photographs which, in mining companies' reports, tend to depict mine sites, workers dressed in safety gear using equipment, and community members undertaking their usual lives or participating in company-run programs. Often, social issues are presented through case studies (Blowfield 2007), as opposed to quantified graphs, detailing the experiences of individuals or community groups, relative to the company's activities.[8] Reports may be guided by a range

---

7   Unlike annual financial reporting where performance indicators are legally defined through financial auditing requirements, the indicators used in sustainability reports are most often part of voluntary reporting initiatives or internal, company policy documents.

8   Interestingly, Blowfield (2007) argues that the reliance on case studies in sustainability reports has the effect of discouraging companies from undertaking more rigorous assessments.

of voluntary principles or indicators, and in some instances, companies may choose to have their performance against these indicators accredited by an external assurance agency. With this description of sustainability reports in mind, we can now think about how best to tap into the information they offer, particularly in relation to gender issues.

Method: Content and Discourse Analysis

This study adopts content and discourse analysis methods to explore the ways and extent to which gender issues are presented in Australia-based mining companies' sustainability reports. The combination of quantitative content analysis with qualitative discourse analysis provides a deeper understanding of the reports' focus and substance. Importantly, both content and discourse analyses focus on text as a site for creating socio-political meaning and revealing agendas and priorities (Eels 1956; Esterberg 2002; Neuendorf 2002). These methods are therefore used together to explore which gender issues Australia-based mining companies include under the mantle of sustainable development, and to unpack the narratives being used to communicate about these issues.

Purposive Sample

This study examined a total of 18 sustainability reports produced by five leading Australia-based mining companies between 2004 and 2007. The sample is not random, nor does it attempt to be wholly generalisable. Instead, a purposive sample was used to construct a data set which allows exploration of sustainability reporting patterns amongst leading Australia-based mining companies. A purposive sample is the most appropriate sample type for this study as it facilitates construction of a data set which meets the specific needs and aims of the research (Neuendorf 2002).

The purposive sample was chosen based on several criteria aimed at creating a data set of mining sustainability reports which represent current reporting practice for Australia and which have comparable sustainable development programs. All companies studied were classified as 'major players' by the 2007 IbisWorld 'Mining in Australia: Industry Report', which is a key summary of business indicators for the mining industry.[9] All companies are either listed

---

9 I have purposefully chosen not to identify the companies analysed in this research. This is primarily related to the aim of the research. The objective here is to demonstrate what leading Australia-based mining companies are reporting about gender and how gender is narrativised throughout these reports. Naming the companies studied would likely result in politicisation of the research, shifting the focus away from the central issue of gender and onto concern over which companies have said what.

on the Australian Stock Exchange (ASX) and/or have regional headquarters and major operations within Australia and the Pacific, meaning they share similar investment in the Australian economy and operate within similar socio-geographic contexts. Since studies have shown that a company's size, location and ownership structure affect its ability to produce a sustainability report, with 'cost and resource constraints' cited as the major impediments to report production (Australian Department of the Environment and Heritage 2005), the companies chosen are also those most likely to be able to consistently invest in sustainability reporting now and into the future.[10]

The 18 sustainability reports which make up the data set included four reports from four companies (a subtotal of 16 reports) and another two reports from a fifth company (total data set of 18 reports). In the case of the fifth company, only two reports were available, as the selected company stopped publishing sustainability reports in 2006, moving instead to production of an annual 'sustainability report' website. The design and technical issues presented by the annual sustainability website meant that data for this company from 2006 and 2007 could not be comparably analysed against traditional reports. To maintain consistency of the data set, it was decided to forego the 2006 and 2007 data from this company.

Most companies produce both 'full' and 'summary' sustainability reports. Where full sustainability reports were significantly longer than 100 pages, and where summary reports were available, summary reports were coded in the interest of time constraints and to forestall 'coder fatigue' (Neuendorf 2002).[11] Summary reports present the key sustainability issues, as defined by companies, and therefore also offer insight into prioritisation of issues. In total, the content analysis examined 404 957 words or approximately 1157 pages of sustainability reporting. All content identified as referring to gender issues was then analysed through a qualitative discourse analysis.

Content Analysis

Content analysis is an exceptionally broad method, incorporating numerous possible approaches, including statistical, computer-automated, network mapping and linguistic techniques (Markoff et al. 2008). For the purposes of

---

10   The profitability and size of Australia-based mining companies accounts, in part, for the fact that 55 per cent of sustainability reports produced in Australia come from the mining and manufacturing sectors (Department of the Environment and Heritage 2005).

11   In several instances, full sustainability reports were over 500 A4 pages long, making them unwieldy for content analysis. In fact, a recent KPMG International and SustainAbility (2008) survey found that of the majority public who do not read sustainability reports, 35 per cent indicate this is due to the reports being too lengthy or reporting websites to difficult to navigate.

this research, content analysis is defined as 'any research technique for making inference by systematically and objectively identifying specified characteristics within text' (Stone et. al. quoted in Markoff et al. 2008: 270). Despite varied definitions, most social scientists would agree that a content analysis is quantitative, 'methodical', and 'systematic', and employs specifically defined and carefully followed rules, such as that all text must be coded (Markoff ibid.).

In brief, content analysis usually consists of the researcher choosing a specific text or set of texts and creating a 'content analysis dictionary' or 'coding frame' through which to read and analyse those texts (Stone et al. 2008). The 'language signs' named in the coding frame may represent phrases or ideas (thematic content analysis) or a particular word (textual content analysis) (ibid.). This study employs both thematic and textual content analyses to provide a thorough reading of the sustainability reports. The textual content analysis included the word 'women' or variations thereof, such as 'woman', 'female', 'females', 'girl', and 'girls'.

In relation to the thematic content analysis, the study explores both 'manifest' (readily observed themes consistently communicated in similar language) and 'latent' themes (Neuendorf 2002). The latent themes represent 'unobserved concepts that cannot be measured directly but can be represented or measured by one or more indicators' (Hair et. al. quoted in Neuendorf 2002: 23). Latent themes are particularly important for this analysis, as they allow intangible issues such as community-based gender issues to be coded. Definitions for both manifest and latent themes employed in the analysis are available in the coding frame (see Appendix 9-1).

Great effort was made to ensure that both thematic and textual code definitions have internal consistency and are mutually exclusive, and that data could be coded consistently across reports (Neuendorf 2002). An intercoder reliability check was performed on all gender-related codes, with a total percentage agreement of 0.84 for the thematic codes and 0.97 for the textual code. Reliability beyond chance agreement was also checked by using Cohen's kappa. The Cohen's kappa score of 0.72 demonstrates a good level of consistency in the coding frame (ibid.) (see Appendix 9-2 for full details of the intercoder reliability check).[12]

The creation of the coding frame is, in itself, an important part of the analysis. A coding frame must be constructed with 'a view to testing one or more theories' (Stone et al. 2008: 132), and should be mostly complete prior to undertaking the analysis. This a priori creation of the coding frame helps to maintain balance between 'objectivity/intersubjectivity' (Neuendorf 2002: 11). In other words, an a priori coding frame is based on knowledge deducted from prior research,

---

12 My thanks to Dr Kay Cook of Deakin University in Melbourne for donating her time and expertise to test the intercoder reliability of my coding frame.

meaning the codes applied are not inordinately shaped by any one document being analysed. A priori construction of the coding frame does not mean, however, that all codes must be established prior to coding. In some instances, an 'emergent dictionary' of select terms or themes may be established through direct interaction with the texts being studied (ibid: 129).

The coding frame for this analysis was largely developed a priori, with all broad and most intermediate level categories determined prior to reading the reports, and more specific categories defined iteratively through familiarity with the reports' contents. For example, 'environmental issues' was predefined as a broad category within the frame, while the more specific 'product stewardship' was created to reflect the reports' content in greater detail. A priori codes were developed deductively from existing documents and research relevant to this study. These documents included studies of mining's social and environmental impacts (for example Macdonald and Rowland 2002; Banks 2003; Vittori et al. 2006; Smith 2008), corporate social responsibility/sustainable development literature (for example Hilson 2000; Moon et al. 2003; Hopkins 2007; Hutchins et al. 2007) and voluntary reporting initiative frameworks (for example IOS 2004; GRI 2006). Sustainability reports from other industries were also examined to situate the coding frame within relevant sustainable development discourses (see BP 2006; Johnson and Johnson 2006; Chevron 2007; Citigroup 2007; ConocoPhillips 2007).

It is important to note that the gender issues explored here made up an important component of a much broader content analysis of these same sustainability reports.[13] It is not the aim of this chapter, however, to discuss the results of the full content analysis. Instead, I wish to focus primarily on the gendered aspects revealed by application of the coding frame.[14] At times, however, findings from the broader analysis will be employed to illustrate the ways in which gender is located within a broader sustainability discourse. The selective study of gender through content analysis, like that presented here, is also common amongst other studies, especially those examining mass media texts. Indeed, this chapter's focus on gender reflects a trend among users of content analysis methods, with Nuendorf (2002: 201) noting, 'Perhaps no substantive area has been more frequently studied [through content analysis] across all the mass media than that of the roles of males and females'.

---

13   The broader content analysis, of which the analysis of gender issues is an important part, was undertaken for the purposes of a Ph.D. research project (in progress) examining the social effects of mining within a corporate social responsibility framework.

14   The full coding frame identifies sustainability issues, including environmental, social and economic effects of mining. The coding frame also examines categories of stakeholder engagement, employment and 'value added' (see Appendix One).

Discourse Analysis

The chapter also employs discourse analysis methods to provide qualitative insights into the ways in which gender is narrativised through sustainability reports. Discourse analysis is used in a range of fields, from linguistics to cultural studies, to investigate the meanings and narratives constructed through choice of language (Johnstone 2008). The discourse analysis method employed here explores the types of language used to communicate about gender issues. The analysis also examines the structure of the reports, surveying where gender issues are situated within the broader sustainability narrative. Questions such as: 'In what contexts do gender issues appear within the reports?' and 'How is gendered language used?' help to shape the discourse analysis.

Discourse analysis views language as inherently political and as shaping both social understandings and identities (Gee 2005). Dissimilar to content analysis, discourse analysis is 'not an algorithmic [quantitative] procedure'; it is far more tractable and adaptive, and involves a great deal of qualitative interpretation by the researcher (ibid.: 6). It is important, therefore, to recognise that both content and discourse analyses, such as those presented below, are inescapably influenced by the interests, experiences and proclivities of the researcher undertaking the analysis. For example, in the case of the sustainability reports studied here, I bring my professional experiences of working for a non-government organisation on mining issues, as well as further professional experiences working in the field of sustainability, to the analysis. My academic background in women's and gender studies also shapes the way I define and analyse gender issues.

It is acknowledged that my own experiences mould the 'situated meaning' constructed through the discourse analysis and also shape the issues examined through the content analysis. 'Situated meaning' can be defined as the socio-cultural experiences which readers bring with them to texts. Importantly, situated meaning can help to build understandings of particular concepts and phrases (ibid.). Situated meanings are also dynamic and malleable; suggesting that a meaning derived from a particular text at one point in time may be changed at a later date, should the reader bring new or different personal experiences to the reading.

The discourse analysis presented here uses the content analysis as a jumping-off point. References to gender issues or 'women' were identified through the content analysis and surrounding paragraphs were then read to determine the narrative shaping these issues. Photographs, charts, graphs and tables within the vicinity of the text in question were also examined to shape more fully the narrative discourse analysis; that is to say, to ensure that the examined narrative was analysed within its complete context. Texts were read to search for shared and similar narratives within and across reports, to look for changes in expression about gender over time, and to see whether and how discussions of gender might be framed or limited.

Importantly, discourse analysis allows exploration of text that is 'missing', not only the text that is written down. As Johnstone (2002: 20) explains, 'Even written texts of the most prototypical sort are the result of decisions about entextualization based on culture-specific expectations'. Discourse analysis therefore encourages us to consider why the gender narrative is narrow and to theorise about the decisions made off the page which lead to such a restricted narrative. For example, what socio-cultural expectations have influenced the deficiency of gender narratives within mining companies' sustainable development discourse? Through discourse analysis, the meager gender narratives presented offer something even in their limitations.

What are Sustainability Reports Saying about Gender?

The analysis below presents the key findings of the thematic and textual content analysis and the narrative discourse analysis performed on the 18 reports that make up the data set. Results from the content analysis are presented quantitatively, in graphic format, while the narrative discourse analysis is presented as complementary qualitative findings. Overall, the study finds that gender is on the radar for mining companies, but it remains a peripheral sustainability consideration.

Like many other gender-focused studies, this analysis confronts the challenge of differentiating 'gender' from 'women,' in an effort to keep distinct two closely related but separate concepts. This distinction is particularly tricky when examining mining, as the historical male dominance of the industry normalises the male gender to such an extent that that it is easy simply to default to looking for mentions of women as indicators of 'gender.' While much information can be gained from such an analysis, this somewhat easier path hinders more nuanced insights into the ways in which gender—encompassing both men and women—is defined and understood by mining companies. It is equally true, however, that it is often only through a discussion about women that gender issues in mining become visible. For this reason, the gender-focused analysis zooms in on report contents which discuss women or community issues traditionally related to women, such as maternity. It is in these instances where women are made visible, that we gain a clearer picture of the way mining itself is engendered and in which both male and female situations are understood.

Gender Issues at the Periphery of Sustainability Reporting

Returning to the first research question: What are Australia-based mining companies putting into the public domain about gender issues, relative to their operations? Content analysis findings indicate that gender issues are seen as marginal in relation to other major sustainability issues, such as the environment. Indeed, gender ranks among the least mentioned issues, relative to sustainability (see Figure 9-1). Of the total content devoted to key sustainability concerns, gender was the least mentioned issue, comprising only 3.5 per cent of total sustainability issues content, followed by health (6 per cent). Employment (36.4 per cent) and environmental concerns (38.7 per cent) make up the bulk of sustainability issues content.

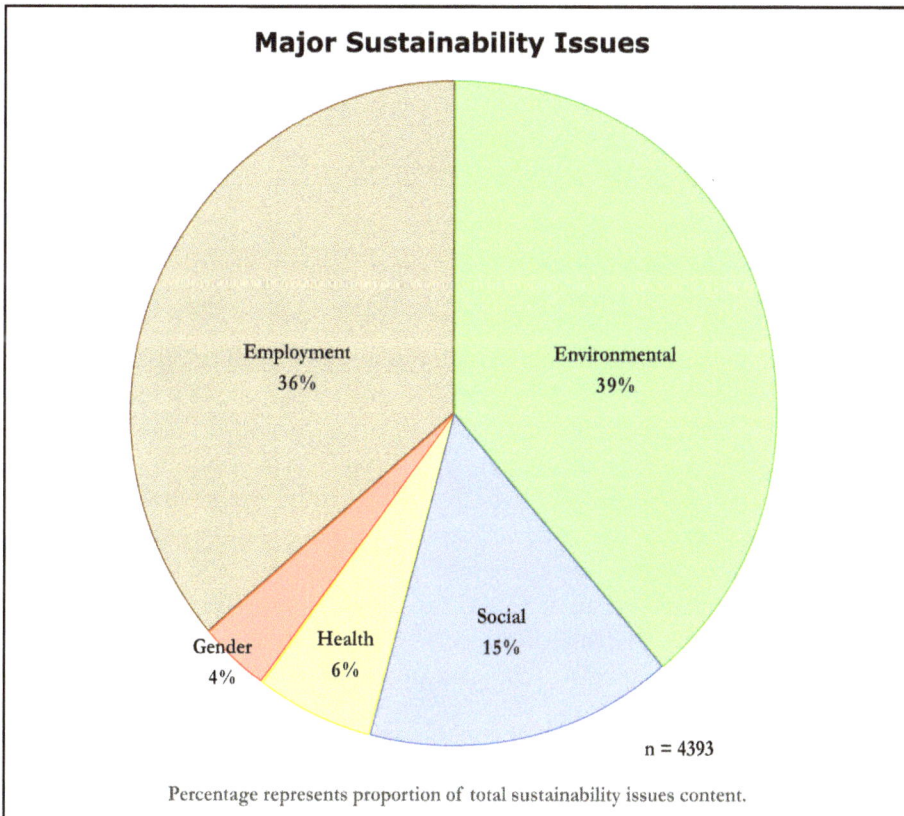

Major Sustainability Issues

Employment 36%
Environmental 39%
Social 15%
Health 6%
Gender 4%

n = 4393

Percentage represents proportion of total sustainability issues content.

Figure 9-1: Major sustainability issues as a proportion of total sustainability issues content.[15]

Source: Author's data.

15   Unless otherwise noted, 'n' represents the total number of mentions of the content analysis codes referred to in each graph.

While a comparison of thematic content, like that shown in Figure 9-1, cannot reveal the importance of issues, it does suggest a prioritisation of issues. That is, those issues which appear most often are also those which are most likely to be prioritised as significant by mining companies and, relationally, their sustainability report audience. Indeed, as sustainability reports are a primary means of stakeholder communication, it can be inferred that their content reflects both those concerns in which mining companies are directly interested, as well as those concerns in which companies believe their stakeholders to be interested (Coombs 2001). Hence, the dearth of gender issues addressed through the reports arguably reflects not only company priorities, but those of target audience—shareholders, potential investors, stakeholders and company employees.[16]

It is also possible that gender content may appear less often than other sustainability issues not because gender issues are socially insignificant, but more so because the language being used by companies is generally neutered. Throughout the reports, issues, such as HIV/AIDS, which would likely take on a gendered narrative in other contexts, are presented in gender-neutral terms. For example, in a 2005 report by a company with African operations, HIV/AIDS is discussed in business-like language, saying in part:

> We train peer educators who run HIV/AIDS education programs, distribute and promote the use of condoms and provide referrals for voluntary HIV counselling and testing (Company B, Report no.6, 2005, 28).

Such neutral language is the norm throughout the sustainability reports studied. Indeed, as Heledd notes in a 2004 study of CSR in mining, companies tend to steer towards more '"inclusive" language, such as using the word "community" to describe a diverse range of stakeholders, embracing local communities, employees and aboriginal groups' (Heledd 2004: 28). While by no means inaccurate, this style of language masks the gendered nature of many sustainability issues. It also raises the question of why these issues are being presented in a neutered manner when so many sustainability issues arguably pose distinctly gendered effects? Possible answers to this question are suggested later in the chapter when analysing the indicators used to guide sustainability reporting.

When language within the reports was gendered, the conceptualisation of gender was quite narrow. Only a few, relatively broad, gender issues were addressed through language which evoked vague thematic ideas, such as 'diversity,' 'empowerment' or 'equality'. In relation to the content analysis, the lack of specificity used in relation to reporting on gender meant that no emergent codes could be identified; all three thematic codes, 'gender issues', 'gender diversity' and 'Equal Employment Opportunity' (see Appendix 9-1 for definitions) and

---

16   A 2006 SustainAbility and UNEP study found that many companies write sustainability reports primarily for investors, while others anticipate their use by employees and affected stakeholders within the community.

the one textual code ('women') were defined prior to engaging with the text. Conversely, environmental concerns were written about much more specifically and in far greater depth, allowing for identification of 12 distinct environmental issues codes, including both a priori and emergent codes (see Figure 9-2).

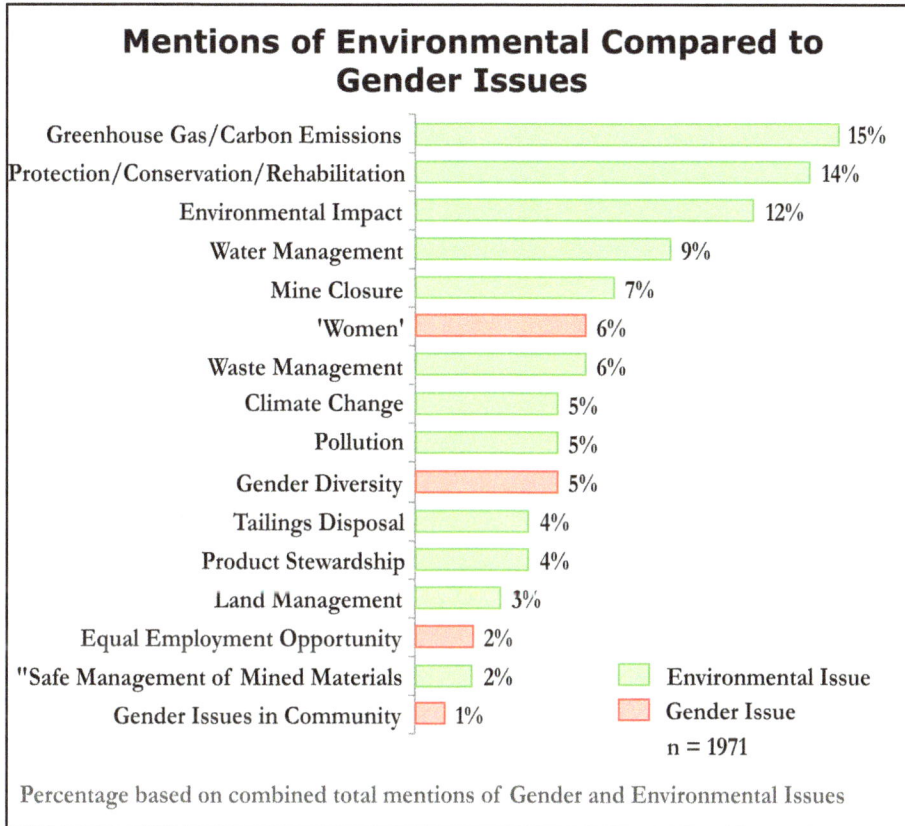

Mentions of Environmental Compared to Gender Issues

Issue	Percentage
Greenhouse Gas/Carbon Emissions	15%
Protection/Conservation/Rehabilitation	14%
Environmental Impact	12%
Water Management	9%
Mine Closure	7%
'Women'	6%
Waste Management	6%
Climate Change	5%
Pollution	5%
Gender Diversity	5%
Tailings Disposal	4%
Product Stewardship	4%
Land Management	3%
Equal Employment Opportunity	2%
"Safe Management of Mined Materials	2%
Gender Issues in Community	1%

Environmental Issue
Gender Issue
n = 1971

Percentage based on combined total mentions of Gender and Environmental Issues

Figure 9-2: Environmental issues compared to gender issues as a proportion of content.

Source: Author's data

The depth to which other sustainability issues are discussed, in comparison with gender, is perhaps best illustrated through a comparison of reporting on 'gender issues' and 'environmental issues.' As shown in Figure 9-2, a range of environmental issues were discussed in the reports, with many specific issues, such as 'mine closure,' and 'product stewardship' regularly addressed across companies and years. Reports generally devote a significant proportion of text to environmental issues, presenting them in their own chapter or section. Gender, in contrast, appears much more sporadically, with only one 2006 report having a dedicated section entitled, 'Gender-Inclusive Community Development' (Company E, Report 3).

The low level of reporting on gender issues appears symptomatic of a broader trend in sustainability reporting, whereby social issues are generally given lesser priority than environmental issues (GRI, HKU and CSR Asia 2008). As Blowfield (2007: 689) notes, while companies of all industries continue to expand efforts to address sustainability concerns, 'we know much more about environmental impact than social or economic impact.' Figure 9-3 shows this trend is apparent among the mining companies studied, with 'environmental issues' 2.5 times more likely to be reported on than 'social issues'. This is despite the fact that a recent SustainAbility report found that the extractives industry rates the importance of social and environmental issues as being of almost equal priority (SustainAbility and UNEP 2006). It is important therefore, to begin to investigate why such a distinct gap currently exists between companies' stated sustainability priorities and their activities.

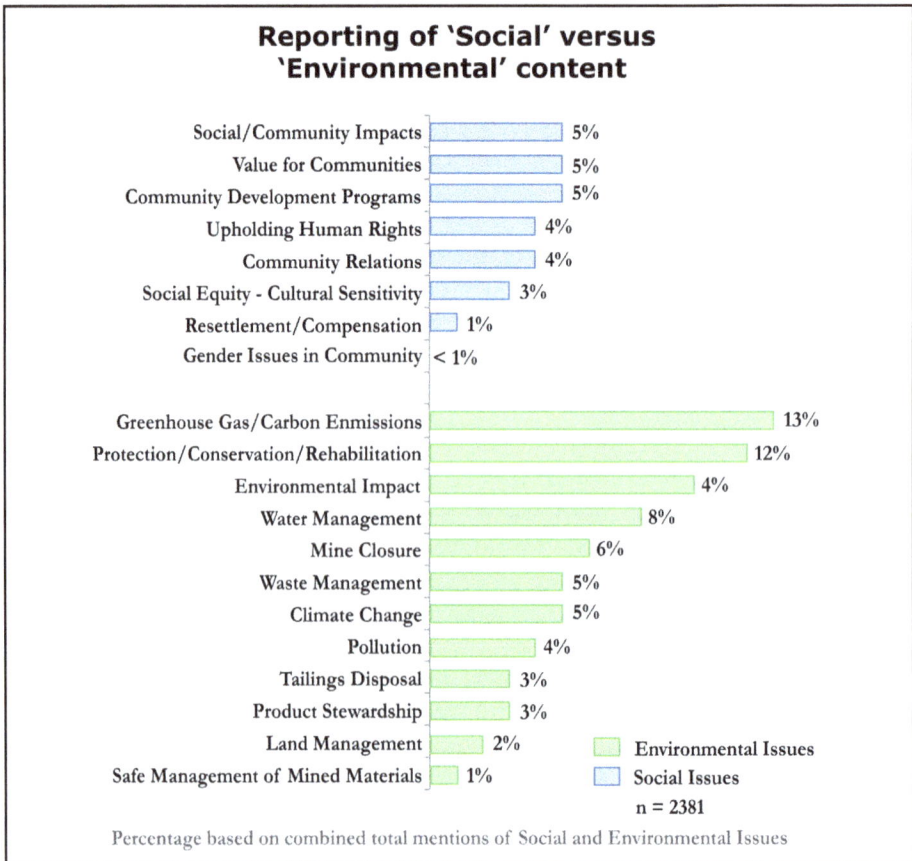

Reporting of 'Social' versus 'Environmental' content

Social Issues	%
Social/Community Impacts	5%
Value for Communities	5%
Community Development Programs	5%
Upholding Human Rights	4%
Community Relations	4%
Social Equity - Cultural Sensitivity	3%
Resettlement/Compensation	1%
Gender Issues in Community	< 1%

Environmental Issues	%
Greenhouse Gas/Carbon Enmissions	13%
Protection/Conservation/Rehabilitation	12%
Environmental Impact	4%
Water Management	8%
Mine Closure	6%
Waste Management	5%
Climate Change	5%
Pollution	4%
Tailings Disposal	3%
Product Stewardship	3%
Land Management	2%
Safe Management of Mined Materials	1%

Environmental Issues
Social Issues
n = 2381

Percentage based on combined total mentions of Social and Environmental Issues

Figure 9-3: Content on 'social issues' compared to 'environmental issues'

Source: Author's data.

Figure 9-3 also shows that companies rarely consider gender issues in amongst other social issues, with 'gender issues in community' comprising less than one per cent of all social issues content covered. The majority of thematic content concerning gender appears instead under the guise of employment issues. Indeed, throughout all the reports studied, gender was most likely to be spoken of in relation to employment/employee issues, particularly in regard to Equal Employment Opportunity (EEO). In some instances, related gendered employment policies and practices, such as maternity leave provisions or flexi-time options, were also included in the discussion. The discourse analysis further reveals the ways in which gender was most frequently narrativised throughout companies' sustainability reports. The following one-sentence reference to an EEO policy epitomises the most common language style use to convey gender concerns:

> The distribution of our employees by gender at 31 December 2005 reflects the effectiveness of our equal opportunity and employment policy in what has traditionally been a male-only industry (Company E, Report 6, 2005).

Most often, statements like the one above are accompanied by charts or graphs which showed a breakdown of the companies' employees by gender. In some instances, effort is made to communicate that the company is making proactive efforts to increase the number of women employees, not simply overall, but specifically in management and senior management roles. Interestingly, when photographs of women employees are used, they rarely depict women doing 'dirty' mining work, but instead show them undertaking environmental studies in pristine outdoor settings or typing at computers. This is in contrast to the majority of photos in the reports which primarily depict men in hard hats, often operating large machinery or posing with an open pit in the background. The narrative around working to increase the number of women employees was present across the four years, but this topic was given greater attention in more recent years (see Figure 9-4).

Significantly, although gender was most often discussed within an 'employment' narrative, gender issues represent only a small portion (five per cent) of total employment issues reported (see Figure 9-5). Again, it is clear that although gender is on the agenda of Australia-based mining companies' sustainability programs, it is a consideration that remains at the periphery.

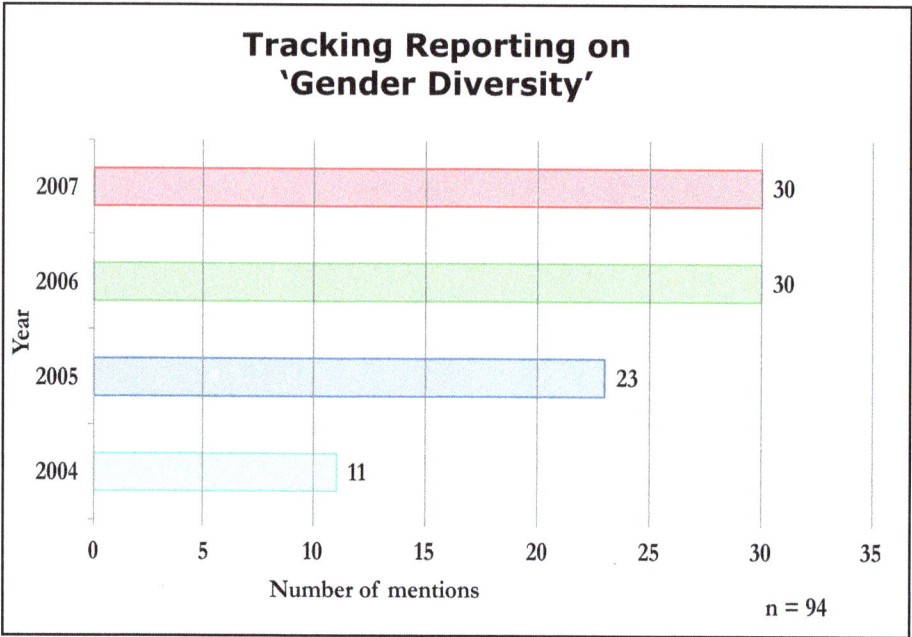

Figure 9-4: Gender diversity reporting trends: 2004–2007.

Source: Author's data.

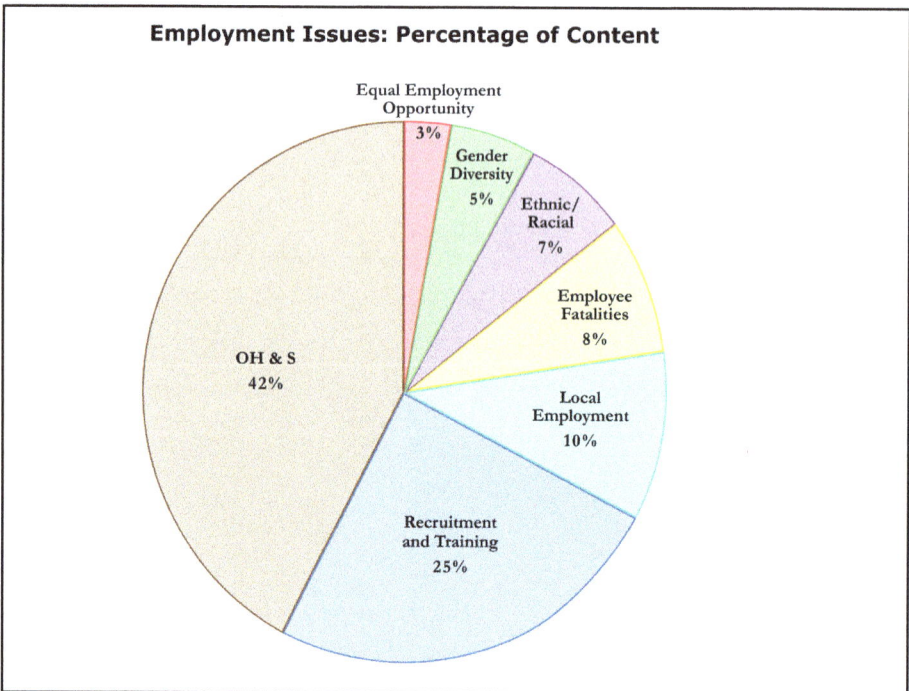

Figure 9-5: Types of employment issues discussed as a proportion of content.

Source: Author's data.

The analysis was able to investigate trends in issues reported, by comparing reports produced over the four-year period. The reports did show that the occurrence of reporting on gender issues has quantifiably increased over the studied period. For example, reporting on 'gender diversity' (see Figure 9-4), has consistently increased since 2004, but gender still remains amongst the least talked about sustainability issues. Again, while a content analysis of this type does not allow for inferences of significance, it seems realistic to assume the consistently increasing number of mentions of 'gender diversity', over the years reflects an improved awareness, prioritisation or concern for gender issues.

So, what does this information tell us about the ways in which companies are approaching, prioritising and engaging with gender issues? In summary, the content and discourse analyses show that gender is only marginally considered as a sustainability issue. Gender rarely appears in discussions of broader social or environmental issues, and where gender concerns are presented, they are given limited attention and appear mostly within an employment context. This minor gender narrative suggests that there is a lack of understanding of the ways in which other sustainability issues, such as resettlement and compensation, present gender-specific concerns. Significantly, the circumscription of gender issues to employment concerns reflects both regulations (for example EEO) and voluntary reporting frameworks. The ways in which these frameworks may be influencing companies' gendered considerations are now discussed in greater detail.

Gender Issues at the Edge of the Screen

The content and discourse analysis findings lead us to the final, but vital research question: Why might companies address gender issues in certain ways? Furthermore, why in an era when intergovernmental agencies prioritise gender mainstreaming and when gender equality is listed as one of eight Millennium Development Goals, is gender still at the periphery of mining's sustainable development agenda?

It is widely accepted that the social issues addressed through CSR programs (in all industries) are limited and do not necessarily reflect the breadth or depth of a company's impact on communities (GRI, HKU and CSR Asia 2008). Limitations concerning which social issues are addressed by corporations, including mining companies, seem to occur in the first instance because firms generally lack a full awareness of their social/community impacts, including gender impacts (ibid.). This scarcity of full social impact awareness is likely the case for several reasons, including: lack of evidence-based social impact data (ibid.); difficulty applying existing means to collect such data (Blowfield 2007); historical patterns of company-community interactions; misdirected measurement tools; reactive responses based on community or NGO complaints (Genasci and Pray 2008); and focus on direct philanthropic giving (Mitnick 2000; Blowfield 2007; Blowfield and Murray 2008). As Blowfield (2007: 686) sums up, 'companies frequently fail to identify their main impacts, so that the most significant consequences of the company's operations are often not the ones given greatest attention'.[17]

While all of the above reasons provide plausible explanations for why companies generally fail to report on or thoroughly identify gender issues, there may be another crucial reason. All of the companies studied use the GRI, as well as other voluntary reporting initiatives, to guide their sustainability reporting. For example, all companies also use the ISO 14001 standards for environmental management.[18] In more recent years, companies refer to or are signatories of the UN Global Compact. Companies report against the Millennium Development Goals, International Council of Mining and Metals Principles, and the Dow Jones Sustainability Index survey, among other accountability frameworks. Each of these voluntary initiatives offers companies guidelines or performance indicators against which to benchmark their sustainability performance.

Clearly, companies' usage of these initiatives raises the quintessential 'chicken or egg' question: Which comes first, sustainability programs to address identified social needs which are then reported on via indicators, or sustainability indicators which tell companies what social concerns they will need to address to comply with reporting guidelines? As Power (1997) argues in *The Audit Society*, regulatory frameworks frequently have the unintended effect of shaping the very programs they seek to measure and hold accountable. Like the 'chicken or egg' question, it is impossible to determine with any certainty the extent to which indicators shape the focus of mining companies' sustainable development programs (and, by extension their sustainability reporting), but it is also highly plausible that these indicators are influencing which issues companies prioritise and address.

Along these lines, an examination of the indicators used in the reports studied here reveals a dearth of indicators focused on gender. Firstly, most companies base their sustainability work around internal, 'Health, Safety, Environment and Community' (HSEC) indicators. A review of publicly available HSEC

---

17 Blowfield's (2007) article focuses on Multinational Corporations (MNCs) operating in developing countries. His comments, however, bear insights which are particularly relevant to the mining industry.
18 Companies use these reporting initiatives to differing degrees. Some comply with certain standards, such as the ISO 14001, and have their reports independently accredited. Others maintain that they have undertaken internal auditing and that they have referred to indicators to guide their report writing.

frameworks reveals that many of HSEC documents do not explicitly incorporate gender indicators. None of the studied companies' publicly available HSEC documents specify gender amongst their indicators. These company based indicators seem to confirm a narrow understanding of gender and the way gender issues affect and are affected by broader sustainability concerns.

Secondly, all of the sustainability reports analysed use GRI indicators to some degree,[19] often in combination with other voluntary initiatives. Following from Power (1997), it seems that these initiatives may very well be influencing not only what is reported on but the types of programs that are implemented. A brief examination of the GRI's indicators, in relation to gender issues, suggests that the reporting frameworks used by mining companies are, indeed, shaping their sustainability programs.

The content analysis suggested that reporting on gender issues, particularly beyond the realm of employment, is weaker than reporting on other key sustainability issues, such as climate change or occupational health and safety. While the discursive concentration of gender issues around employment of women could be seen simply as reflecting a broader social trend, in Australia there has generally been a cultural push to increase the number of women on boards and in senior management, with women currently comprising less than 11 per cent of senior executive roles in ASX200-listed companies, and only 3.4 per cent of total board membership amongst all ASX-listed companies (Sheridan 2001; Braund and Medd 2008)—it also appears very likely that the tendency to report on gender in relation to employment is influenced significantly by GRI performance indicators. In the GRI G3 Guidelines (2006), for example, the only mentions of gender or women appear in indicators which fall under the 'Labour Practices and Decent Work' guidelines.

The GRI is extremely limited on gender indicators, with the three gender-specific indicators appearing under the 'Labour Practices and Decent Work' category (GRI 2008). These include indicators such as, 'Ratio of basic salary of men to women by employee category' (ibid.: Indicator LA14). The trend for ensconcing gender issues within labour/employment issues in the GRI indicators seems to be directly reflected in the ways in which the studied mining companies are approaching gender.

This begs the question, are gender issues just a blip on the periphery of the sustainability radar? Recent action by the GRI suggests gender issues, especially those linked directly to women's experiences, are becoming

---

19   The GRI is a voluntary reporting framework which 'provides guidance for organizations to use as the basis for their disclosure about their sustainability performance' (see GRI 2007). Corporations may choose to use different 'Application levels', which indicate the degree to which the indicators were applied, and whether external, independent validation of the report was conducted.

increasingly visible. Indeed, a current initiative by the GRI and related groups to address gender issues may help to forestall the negative effects of neglecting gender, relative to sustainable development programs. In September 2008, the GRI admitted shortcomings of its indicators in relation to gender, announcing a 12-month joint research project with the International Finance Council to develop a 'Gender Sustainability Reporting Guide' (GRI 2008). While only just beginning, this research marks an important step in more comprehensive incorporation of gender issues into sustainability reporting frameworks, and it will be interesting to see whether the inclusion of more gender indicators in these frameworks eventually leads to improved gender-sensitivity in mining companies' sustainable development programs.

Specifically in relation to mining, the GRI, in consultation with a multi-stakeholder group from the International Council on Mining and Metals (ICMM), has developed pilot supplementary indicators for the Mining and Metals Sector (MMS) (GRI 2005). The 2005 pilot supplement included one mention of 'women', in relation to 'community engagement,' positioning them as a 'marginalised' group requiring attention. The proposed engagement with women detailed in the pilot MMS supplement did not constitute a measurable performance indicator, however, but instead appears under the section, 'Aspects to be Reported through Narrative Description' (ibid.).

In early 2009, the GRI released a revised draft MMS supplement. Here, engagement with women has been incorporated within performance indicator SO1: 'Nature, scope and effectiveness of any programs and practices that assess and manage the impacts of operations on communities, including entering, operating and exiting' (GRI 2009). While gender issues are not explicitly raised within the draft indicator, the 'MMS Commentary' to guide reporting against this indicator suggests that:

> Reporting on the extent to which community participation processes are socially inclusive and which ensure engagement with women or disadvantaged groups can indicate a company's ability to manage relationships with groups that otherwise might not benefit from its operations (a social responsibility aspect) or which can oppose operations if they do not feel they are treated equitably (a risk mitigation aspect) (ibid.: 55).

Ultimately, it is vital that awareness of gender issues be stretched beyond employment concerns to incorporate those less tangible but still very real gender impacts that mining operations may bring to communities of operation. This may occur through refinement of commonly used reporting frameworks, such as the GRI, to better incorporate gender issues. Refined performance indicators would better focus companies' attention on gender issues, providing direction as to where to begin when addressing gender concerns. While the GRI's shift towards incorporating 'women' within the description of a community engagement indicator, for example, is an important first step, it does not go far enough. As new indicators are developed, it is vital that gender-specific indicators are developed. The alternative is that gender issues will continue to be seen as an immeasurable or 'soft issue,' and therefore, arguably given less priority than other issues with their own indicators. Similarly, company HSEC policies need to be refined to better incorporate gender issues within their performance indicators.

Although only a cursory insight requiring further examination, it seems fair to suggest that voluntary reporting indicators are, indeed, shaping the ways in which companies are thinking about, incorporating and reporting on gender issues within their sustainable development programs. Future research might use interviews with mining companies to determine the extent to which they feel indicators shape sustainable development programs and reporting, and vice versa. These insights would begin to answer the 'chicken or egg' question, and, if data suggests that indicators do indeed shape sustainable development programs, more concerted effort could be made to create indicators which work to measure the complete scope of mining's effects.

While indicators play an important role in influencing the ways in which sustainability issues are prioritised, production of gender indicators is not the only answer. Companies themselves must be encouraged to address gender issues, and communities must be aided in articulating the gender-specific effects of mining. It is also suggested, therefore, that development of a systematic, reliable and valid means of measuring gender impacts in mining communities, beyond the realm of employment, would improve awareness of the depth and breadth of these issues. Such data could then be used to plan and implement targeted sustainability programs, bringing gender onto the radar.

Appendix 9-1: Thematic and Textual Content Analysis Coding Frame

Thematic Codes		
Broad Level Category	Intermediate Level Category	Definition
Sustainable Development	Sustainable Development Indicators	includes mentions of key performance indicators, sustainable development/sustainability targets, and measurement of sustainable development
Corporate Social Responsibility	Business case	includes arguments for why and how sustainable development is 'good for business' or provides shareholder, stakeholder or business value. Includes text where direct linkages are made between sustainable practices and financial performance. Includes rationale for investing in sustainable development, where this rationale asserts that 'business' or 'financial' value has been gained for the company
Environment	Pollution	includes air pollution, spills and accidents where chemicals or tailings have affected the local environment
	Safe management of mined materials	includes mentions of the ways mined materials, particularly uranium, are handled, particularly to prevent pollution, hazards or damage to the local environment
	Product stewardship	includes all mentions of 'product stewardship', efficient use of resources, supply chain management, procurement, product life cycle and related concepts of caring for the product from exploration through extraction phases to maximise value
	Water management	includes mentions of storm water drainage, water management, drainage, run-off, and policies and procedures in place to manage these
	Land management	includes mentions of land management, land use, policies and protocols referring to environmental aspects of land use, but not 'land use agreements' or 'land use negotiations'
	Waste management	includes mentions of 'waste management', 'waste disposal', 'hazardous waste disposal', 'hazardous waste', including policies and procedures to deal with waste and protocols for how waste is stored
	Environmental impact	includes mentions of 'environmental impact', 'environmental impacts', including environmental impact assessments, environmental impact mitigation plans, reducing environmental impact and managing environmental risks

	Protection/ Conservation	includes mentions about protecting or conserving biodiversity and the ecosystem; includes 'environmental stewardship'
Community/ Social	Social/Community Impact	includes mentions of 'social impact', 'social impacts', including social impact assessments, social impact mitigation plans, reducing social impact and managing social risks
	Social equity/ Cultural sensitivity	includes mentions of cultural awareness or expressions of sensitivity to local customs; treating others within the community as equals and with respect
	Community relations	includes mentions of 'community relations', including CR programs, policies and staff. Also includes mentions of 'community relations plans'
	Resettlement/ Compensation	includes mentions of 'resettlement', including resettlement plans, resettlement experiences and negotiations, as well as related experiences of compensation, including compensation negotiations and tracking how compensation has been used in the community
	Upholding human rights	includes mentions of 'human rights', 'people's rights', 'rights', human rights doctrines or protocols, examples of efforts made to uphold human rights, human rights standards, mentions of averting human rights violations
	Value for communities	includes economic development or improvements in economic status which affect local communities; poverty reduction; 'sustainable livelihoods'
	Community development	includes mentions of 'community development', as well as discussion of development plans, related community development outcomes, 'sustainable communities', 'capacity building' and 'development' within a community or socio-cultural context
	Gender issues in community	Includes non-employment related gendered issues. For example, women in the community, women's health concerns (including maternal health), family management (including child care), female headed households, consultation with women in community. Mentions of issues, such as HIV/AIDS, which affect women but which are not discussed in specifically gendered terms, are not included under this code.
Community Health	Community Health Education	includes mentions of community health education programs sponsored or run by the mine, related case studies and examples, community health education also includes health education programs run for employees, such as HIV/AIDS programs

	Investment in/ Support for community health programs	includes company donations or sponsorship of community health programs or infrastructure, such as building health care centres or funding a doctor's position, includes support for community health programs such as immunisation programs
Employment/ Employees	Employee fatalities	includes mentions of 'employee fatalities', 'fatalities', 'employee deaths' and steps taken or policies implemented to reduce the number of employee deaths in future. Also includes condolences to employees' families
	Recruitment and training	includes mentions of recruitment processes, including transparency in recruitment, training and professional development programs for employees and also for members of the community, such as in-community skills programs and apprenticeships sponsored by the company
	Volunteering	includes mentions of employee volunteering programs and examples of volunteering undertaken by employees within the community, also includes accounts of the number of employee volunteer hours donated
	Gender diversity	Includes all mentions of gender equity or gender diversity appearing outside of Equal Employment Opportunity statements. For example, mentions of attempts to increase the number of women on boards of management, gender diversity targets, commitment to a gender-diverse workforce or mentions of 'women in mining'.
	Equal Employment Opportunity	Includes all direct mentions of 'Equal Employment Opportunity' clauses and policies. Includes policies on non-discrimination, child care, maternity leave, family leave, fair and equitable treatment of women or prevention of sexual harassment.
	Local employment	includes policies to employ people from communities adjacent to the mine site, percentages and numbers of locals employed at the mine; also includes mentions of how the mine affects local employment in other local industries (e.g. shops which open to serve the influx of mine workers)
	Occupational Health and Safety	includes mentions of 'occupational health and safety', 'OHS', 'Oc-Health', OHS policies and procedures and performance against OHS standards, as well as OHS training
Accountability	Risk management	includes all mentions of risk management, risk assessment and related policies, including actions described as being taken directly for the purposes of risk management, which are not environmental or social impact assessments

	Internal Governance Standards	includes all mentions of internal governance policies which the company uses as accountability measures, such as Health Safety Environment and Community (HSEC) policies determined by the Board, but not Equal Employment Opportunity policies
	Socio-cultural expectations	includes mentions of companies' attempts to meet the social or cultural expectations of the local community where mining is occurring, may also refer to the company's efforts to meet socio-cultural expectations of shareholders or stakeholders more broadly
Accountability	Social impact measurement	includes all references to the specific ways in which companies measure social impacts, but not risk management. The code aims to capture measurement practices, not mentions of 'social impact assessment'.
	Other accountability programs	includes mentions of other accountability frameworks used by companies to guide sustainability reporting and which are not covered under the other codes
	Government regulations	includes all legislation, statutes and other government regulations
	transparency	includes all mentions of 'transparency', 'maintaining transparency', 'operating transparently' or relevant descriptions of practices or policies which imply transparency but which are not related to the EITI
	Environmental impact measurement	includes all references to the specific ways in which companies measure environmental impacts, but not risk management. The code aims to capture measurement practices, not mentions of 'environmental impact assessment'.
Stakeholder engagement		All codes under stakeholder engagement refer to specific relationships and include mentions of the companies' interactions with that particular stakeholder
	Private-public partnerships	includes a company's collaborative relationships with universities, or similar bodies
	Non-governmental/ Intergovernmental agencies	
	National governments	
	Private-private partnerships	includes partnerships with other companies, across any industry
	Local governments	includes both state and local government relationships
	Contractors	
	Shareholders	
	Local communities	

Financial/ Economic	Charitable giving	includes all mentions of donations, philanthropy and gifts to charity
	Value for shareholders	includes mentions of 'value for shareholders', 'shareholder value', and related discussions of how the company is concerned with shareholders and creation of returns
	Sustainable business	includes mentions which use the phrase 'sustainable business' or 'business sustainability' and references to the ways in which the business practices to ensure profit-making and support operations in the 'long-term', 'into the future' or 'over time'
	Investment in environment	includes all mentions which discuss direct financial investment into the environment, including investment into conservation or biodiversity programs, 'environmental expenditure', and investment in technologies to reduce negative environmental impacts.
	Value for governments	mentions of the ways in which the companies operations directly benefit the countries of operation, including tax revenues, support for industry and employment levels and related 'poverty reduction'
	Investment in communities	includes all financial investment in communities which is not a community program or charitable giving. 'Investment in communities' may include construction of community infrastructure, such as sports stadiums or community halls, but not health centres or sanitation/environmental/ transport infrastructure
Textual codes		
Sustainable development/ Sustainability		includes all mentions of 'sustainable development,' 'sustainability'
CSR		includes all mentions of 'corporate social responsibility,' 'CSR', 'social responsibility', 'SR', 'financial, social and environmental responsibility', 'social, environmental and economic performance'
	Ethical practice	includes all mentions of 'ethics', 'ethical', 'ethical practice', 'ethical values', 'ethical business practice', 'ethical conduct', 'business ethics'
	Triple bottom line	includes all mentions of 'triple bottom line', 'quadruple bottom line',
	Licence to operate	includes all mentions of 'licence to operate' or 'social licence'
Environment	Tailings disposal	includes all mentions of 'tailings', 'mine tailings', 'tailings disposal', 'tailings dam', 'tailings storage', 'sediment disposal', 'sediment control', 'acid mine drainage'

	Mine closure	includes all mentions of 'mine closure', 'closure', 'abandonment', 'abandoned mines', 'closure plans', 'mine closure plans', 'rehabilitation and closure'
	Greenhouse gas/ Carbon emissions	includes all mentions of 'greenhouse gas', 'greenhouse gases', 'carbon emissions', 'carbon', 'CO2', 'carbon output', 'gas emissions', and 'emissions'
	Climate change	includes all mentions of 'climate change' or 'global warming'
Globalization		includes all mentions of 'globalization', 'globalising', 'global market', and lists of international sites of operation
Accountability	Global Reporting Initiative	includes all mentions of 'Global Reporting Initiative' or 'GRI'
	UN Global Compact	includes all mentions of 'UN Global Compact', 'United Nations Global Compact', 'Global Compact' or 'UNGC'
	Millennium Development Goals	includes all mentions of 'Millennium Development Goals', 'Millennium Goals', 'MDGs'
	International Council on Mining and Metals	includes all mentions of 'International Council on Mining and Metals', 'ICMM'
	United Nations Universal Declaration on Human Rights	includes all mentions of 'United Nations Universal Declaration on/of Human Rights', 'UN Declaration on/of Human Rights', 'UN Universal Declaration on/of Human Rights', 'UN Human Rights Declaration',
	Corruption/ Bribery	includes all mentions of 'corruption', 'corrupt practice', 'corrupt business practice(s)', 'bribery', 'bribes' and related policies on bribes and corruption
	ISO 14001	includes all mentions of 'ISO 14001', 'ISO'
	Extractive Industries Transparency Initiative	includes all mentions of 'Extractive Industries Transparency Initiative', 'EITI'
Women		Textual code: includes all exact mentions of the following terms: 'women', 'woman', 'female', 'females', 'girl', 'girls'

Appendix 9-2: Intercoder Reliability Calculations

	Thematic Codes				Textual Code
	Gender Diversity	Equal Employment Opportunity	Gender Issues in Community	Totals	Women
Coder A	26	6	6	38	31
Coder B	29	9	7	45	32
Product of Marginals	754	54	42	850	992
Agree	26	6	6	38	31
Disagree	3	3	1	7	1
n =	29	9	7	45	32
Percent Agreement (PAo)	0.90	0.67	0.86	0.84	0.97
Percent Agreement Expected (PAe)	—	—	—	0.42	N/A
Cohen's kappa	—	—	—	0.72	N/A

Formulas for calculation[20]

Percentage Agreement (PA_o) =

$$\frac{Total\ A's}{n}$$

Expected Agreement (PA_E) =

$$\left(\frac{1}{n^2}\right)\left(\sum pm_i\right)$$

Where pm_i = each product of marginals

---

20   Although scholars do not generally agree on how intercoder reliability should be calculated, or what percentage of agreement constitutes good reliability, there is general agreement that intercoder reliability checks should be performed to validate content analyses. The two-coder intercoder reliability performed here measures both simple percentage agreement – that is, the percentage to which both coders were in agreement about their codes - and Cohen's kappa. Cohen's kappa is used to measure the agreement of two coders beyond any 'chance' agreement on thematic codes –that is, to what extent are the codes reliable and replicable when the element of chance agreement is calculated for (See: Neuendorf (2002) for further information).

Product of marginals (pm_i) =

$_n$Coder A codes x $_n$Coder B codes

Cohen's *kappa* (κ) =

$$\frac{(PA_o) - PA_E}{1 - PA_E}$$

References

Atkinson, J. and Community Aid Abroad Australia, 1998. *Undermined: The Impact Of Australian Mining Companies In Developing Countries.* Fitzroy: Community Aid Abroad.

Australian Department of the Environment and Heritage, 2005. *The State of Sustainability Reporting in Australia 2005.* Canberra: Department of the Environment and Heritage. (2nd Edition).

Auty, R.M., 1993. *Sustaining Development in Mineral Economies: The Resource Curse Thesis.* London and New York: Routledge.

Banks, G., 2003. 'Landowner Equity in Papua New Guinea's Minerals Sector: Review And Policy Issues.' *Natural Resources Forum* 27: 223–34.

Blowfield, M., 2007. 'Reasons To Be Cheerful? What We Know About CSR's Impact.' *Third World Quarterly* 28: 683–95.

Blowfield, M. and A. Murray, 2008. 'The Impact of Corporate Responsibility.' In M. Blowfield and A. Murray (eds), *Corporate Responsibility: A Critical Introduction.* Oxford: Oxford University Press.

Braund, C. and R. Medd, 2008. 'WoB Road Map For Gender Diversity On Australian Boards.' Report produced by Women on Boards. Gosford: Women on Boards.

British Petroleum, 2006. 'BP Sustainability Report 2006.' London: British Petroleum.

Chevron, 2007. '2007 Corporate Responsibility Report.' San Ramon: Chevron.

Citigroup, 2007. '2007 Citizenship Report.' New York: Citigroup.

ConocoPhillips, 2007. 'Conocophillips Sustainable Development Report.' The Hague: Conocophillips.

Coombs, T., 2001. 'Interpersonal Communication and Public Relations.' In R. Heath and G. Vasquez (eds) *Handbook Of Public Relations*. Thousand Oaks: Sage.

Eels, R., 1956. *Corporate Giving in a Free Society*. New York: Harper.

Elkington, J., 1997. *Cannibals with Forks: The Triple Bottom Line of 21st Century Business*. Oxford: Capstone.

Emberson-Bain, A., 1994. 'Mining Development in the Pacific? Are we Sustaining the Unsustainable?' In W. Harcourt (ed.), *Feminist Perspectives on Sustainable Development*. London: Zed Books in Association with the Society for International Development.

Esterberg, K.G., 2002. *Qualitative Methods in Social Research*. Boston: McGraw Hill.

Frynas, J.G., 2005. 'The False Developmental Promise of Corporate Social Responsibility: Evidence from Multinational Oil Companies.' *International Affairs* 81: 581–98.

Gee, J.P., 2005. *An Introduction to Discourse Analysis: Theory and Method*. London: Routledge.

Genasci, M. and S. Pray, 2008. 'Extracting Accountability: The Implications of the Resource Curse For CSR Theory and Practice.' *Yale Human Rights and Development Law Journal* 11: 37–58.

GRI (Global Reporting Initiative), 2005. 'Global Reporting Initiative: Mining and Metals Sector Supplement: Pilot Version 1.0.' Amsterdam: Global Reporting Initiative.

———, 2006. 'Global Reporting Initiative: Sustainability Reporting Guidelines.' Amsterdam: Global Reporting Initiative.

———, 2009. 'GRI: Sustainability Reporting Guidelines and Mining and Metals Sector Supplement: Draft Sector Supplement for Public Comment.' Amsterdam: Global Reporting Initiative.

———, 2007. 'Reporting Framework.' Viewed 18 January 2011 at http://www.globalreporting.org/ReportingFramework/

GRI, HKU and CSR Asia (Global Reporting Initiative, University Of Hong Kong and CSR Asia), 2008. *Reporting On Community Impacts*. Amsterdam: Global Reporting Initiative.

Heledd, J., 2004. 'Corporate Social Responsibility and the Mining Industry: Conflicts and Constructs.' *Corporate Social Responsibility and Environmental Management* 11: 23-34.

Hilson, G., 2000. 'Sustainable Development in the Mining Industry: Clarifying the Corporate Perspective.' *Resources Policy* 26: 227–38.

Hopkins, M., 2007. *Corporate Social Responsibility and International Development: Is Business the Solution?* London: Earthscan.

Hutchins, M.J., C.L. Walck, D.P. Sterk and G.A. Campbell, 2007. 'Corporate Social Responsibility: A Unifying Discourse for the Mining Industry?' *Greener Management International* 52: 17–30.

IbisWorld, (2007). 'Mining in Australia: Industry Report' Viewed 9 September 2009 at http://www.ibisworld.com.au/industry/default.aspx?indid=55

ISO (International Organization for Standardization), 2004. *ISO 14001: 2004 Standards Catalogue: Mining And Minerals*. Geneva: ISO.

Johnson and Johnson, 2006. 'Sustainability Report.' New Brunswick: Johnson and Johnson.

Johnstone, B., 2008. *Discourse Analysis*. Malden, MA: Blackwell.

Jones, M., S. Marshall and R. Mitchell, 2007. 'Corporate Social Responsibility and the Management of Labour in Two Australian Mining Industry Companies.' *Corporate Governance: An International Review* 15: 57–67.

Klubock, T.M., 1998. *Contested Communities: Class, Gender and Politics in Chile's El Teniente Copper Mine, 1904-1948*. Durham: Duke University Press.

KPMG International and SustainAbility, 2008. *Count Me In: The Readers' Take On Sustainability Reporting*. London: SustainAbility. (1st Edition).

Kunanayagam, R., 2003. 'Sex Workers: Their Impact On and Interaction with the Mining Industry.' Paper presented at the *Women In Mining* conference, Madang, Papua New Guinea.

Macdonald, I. and C. Rowland, 2002. *Tunnel Vision: Women Mining and Communities*. Presented to the 'Tunnel Vision: Mining, Women and Communities' forum Report. Fitzroy: Oxfam Community Aid Abroad.

Markoff, J., G. Shapiro, G and S. Sweitman, 2008. 'Toward the Integration of Content Analysis and General Methodology.' In R. Franzosi (ed.), *Content Analysis*. Los Angeles and London: Sage.

Mitnick, B., 2000. 'Commitment, Revelation and the Testaments of Belief: The Metrics of Measurement of Corporate Social Performance.' *Business and Society* 39: 419–465.

Moon, J., A. Crane and D. Matten, 2003. 'Can Corporations Be Citizens? Corporate Citizenship as a Metaphor for Business Participation in Society.' *International Centre for Corporate Social Responsibility Research Paper Series*. Nottingham: University of Nottingham.

Neuendorf, K., 2002. *Content Analysis Guidebook*. Thousand Oaks: Sage.

Power, M., 1997. *The Audit Society*. New York: Oxford University Press.

Sheridan, A., 2001. 'A View From The Top: Women on the Boards of Public Companies.' *Corporate Governance* 1: 8–15.

Smith, G.A., 2008. 'An Introduction to Corporate Social Responsibility in the Extractive Industries.' *Yale Human Rights and Development Law Journal* 11: 1–7.

Stone, P., D. Dunphy, M. Smith and D. Ogilvie, 2008. 'The Construction of Categories for Content Analysis Dictionaries.' In R. Franzosi (ed.), *Content Analysis*. Los Angeles and London: Sage.

Storey, K., 2001. 'Fly-In/Fly-Out and Fly-Over: Mining and Regional Development in Western Australia.' *Australian Geographer* 32: 133–48.

SustainAbility and UNEP (United Nations Environment Programme), 2006. *Tomorrow's Value: The Global Reporters 2006 Survey Of Corporate Sustainability Reporting*. London: SustainAbility. (1st Edition)

Vittori, L., S. Martin and S. Bice, 2006. *Mining Ombudsman: Case Updates 2005*. Mining Ombudsman Case Reports. Melbourne: Oxfam Australia.

Walton, G. and J. Barnett, 2008. 'The Ambiguities of "Environmental" Conflict: Insights From the Tolukuma Gold Mine, Papua New Guinea.' *Society & Natural Resources* 21: 1–16.

Wheeler, D. and J. Elkington, 2001. 'The End of the Corporate Environmental Report? Or The Advent Of Cybernetic Sustainability Reporting And Communication.' *Business Strategy and the Environment* 10(1): 1–14.

10. Towards a Post-Conflict Transition: Women and Artisanal Mining in the Democratic Republic of Congo

Rachel Perks

Introduction

This chapter[1] represents a range of experiences by development actors, foreign donors, government and mining companies in response to the challenge of women's issues in the artisanal mining sector of the Democratic Republic of Congo (DRC).[2] Although some statistics and trends are common across the country, this chapter speaks more specifically to the contexts of Katanga Province and Ituri District of Orientale Province. It does not in any way attempt to generalise what is a very diverse economic and social environment. It draws from Pact's[3] work within a public-private partnership aimed at improving governance and livelihoods in the DRC mining sector.[4]

Artisanal and small-scale mining (ASM) in the DRC presents one of the greatest sources of economic opportunity for millions of Congolese citizens. It is estimated that two million people work as artisanal miners across the country, producing 90 per cent of the minerals exported (WB 2008; Hayes 2008). With

---

1 Sections of this chapter have since been used in Hayes and Perks (2011).
2 The learning found within this chapter would not be possible without the concerted efforts of actors to tackle the issues despite the obvious risks of working with such a volatile, illegal and opportunistic sector. In the DRC in particular, Anvil Mining Ltd, Anglo Gold Ashanti and United States Agency for International Development have consistently supported the work done in this domain by the international non-government organisation (NGO) called Pact. Other funding partners have included the International Finance Corporation, Tenke Fungurume Mining, Katanga Mining, and DCP/Nikanor. One cannot ignore the prime role played by Congolese women. In the Congo women face great personal security risks in speaking out against abuse, discrimination and exploitation. The courage to share their stories with Pact staff and researchers is sincerely recognised.
3 Pact Inc. is based in Washington D.C. and has been operating in the DRC since 2003. Pact's DRC country program focuses solely on the responsible management of natural resources which Pact regards as a fundamental requirement for sustaining peace and preventing further conflict. Pact works with private mining companies, local communities and the national government and is funded by a variety of partners.
4 See Appendix 10-1 for Pact's Gender ASM Partners in Katanga Province and Ituri District in Orientale Province.

their dependents, it is estimated that the ASM sector provides economically for 18 per cent of the national population. Although this sector contributes to the livelihoods of such a large proportion of the population, it consistently exhibits some of the worst forms of labour, environmental and social practices found in the DRC today.

As with all recovering conflict countries, the DRC's peace and future for sustainable development rests largely on economic revitalisation. While humanitarian and development aid provides temporary relief to the most vulnerable population groups in an immediate transition period, long-term economic growth depends on outside investment and government regulation of its most profitable industries, with the mining sector being the most obvious.

However, whereas industrial mining will most likely constitute the backbone of the economy in the decades to follow, the real wealth and livelihood for individuals is presently found in the unregulated ASM sector. The transition period from ASM to large-scale industrial mining is wrought with potential conflicts and severe challenges as livelihood opportunities will be lost and access to resources will be restricted.

In the DRC, the current political, economic and social transition from war to peace continues to leave many root causes and drivers of the conflict unresolved. Within this framework of post-conflict transition, the security of communities remains fragile as the pillars of local governance, economic opportunity and social cohesion rebuild at an alarmingly slow pace. Within this void flourishes individual survivalism, impunity, and escapism—primarily by men—into alcohol and drugs.

Gender mainstreaming seeks to examine and redefine the roles attributed to men and women in a given social context. In the DRC's artisanal mining sector, as synonymous with the country as a whole, this is inherently linked to re-building the roles of men and women, and the values of each gender, following a period of conflict. While most advocacy and lobby groups focus on issues of revenue transparency, and the implementation of standards practiced by industrial companies, the deplorable conditions and discrimination facing artisanal miners remains largely overlooked, especially pertaining to women and young girls.

Women constitute between 40–50 per cent of the ASM workforce in Africa (Hinton et al. 2003). They constitute a significant economic force and often carry the burden of providing for whole households. In the DRC, as in most other African countries, women perform support service work to the industry and rarely engage in the physical act of mining. In particular, they also provide the majority of services at the mine sites and adjacent mine camps such as restaurants, markets, and small kiosks. Women work voluntarily and forcibly in the sex trade, often unprotected or under intimidation.

A concerted focus on transformation of the country's ASM sector is critical for ensuring overall stability in the DRC. In this process, gender considerations are imperative. Failure to address these conditions and enhance social capital will most likely result in further social instability, prevent the overall transition from war to peace, and ultimately have the potential to undermine mining companies' 'social license' to operate in such tenuous environments.

The work of Pact and its partners in addressing gender issues in ASM has led to important lessons that can help inform and improve global practice in this sector. Given the sheer number of women artisanal miners and their levels of poverty, overall sustainable transitions and reform of the ASM sector is not possible without targeted support towards gender challenges and issues. However, while gender-focused programming is essential, it must rest within a larger framework of ASM reform in order to ensure buy-in by other ASM actors. Pact's experience has shown that investment in women artisanal miners is more likely to increase broader social capital due to their role as sole income providers in many communities and their relationship to a more traditional and stable rural life. Women constitute a bridge between two very distinct and often isolated development contexts and their role in bringing ASM into the wider development agenda of the DRC is of great value.

Thus, gender issues should be at the forefront of social development initiatives in order to enhance social stability, whether through public-private partnerships, NGO and UN programs, or concession-based mining programs. To ensure this coherence, all actors—including mining companies—should align their development objectives and work within a commonly agreed framework to increase the potential for impact. This is important as the ASM sector has limited funding opportunities and actors in positions to effectively promote gender issues presently. In this regard, private-public partnerships can enhance benefits and limited resources and ensure complementarities.

Background to Artisanal Mining in the DRC

Conflict Aftermath: Continued Importance of ASM

The DRC is a country of great natural resources, mineral wealth and agricultural potential. But a debilitating colonial past and two civil wars have resulted in 30 years of neglect and consequently, at the time of writing: 70 per cent of its

57.5 million inhabitants live below the poverty line; infant and child mortality rates have risen to 92 and 148 deaths per live births respectively; disease is endemic; infrastructure is virtually non-existent; general investor climate is marred by the still-unconcluded Mine Contract Review Process; and by fears of another war resurging in eastern Congo (WHO 2006).

Conflict in the DRC has typically been associated with the drive to gain and maintain control of natural resources. This relationship is not unique to the DRC, as demonstrated in Sierra Leone and Liberia, and is equally at the heart of persistent conflicts in places such as Sudan.

The DRC war from 1999–2003 claimed over five million lives, both directly and indirectly as a result of active fighting. In Ituri District (Orientale Province), it is estimated that 55 000 people died and that another 200–300 000 suffered from severe human rights abuses, often at the hands of fellow villagers and family members (IRC 2004). At the time of writing, 1.5 million internally displaced persons are estimated nationwide with the third and fourth highest concentrations in Ituri District and Katanga Province; 185 500 and 175 815 respectively. While most conflict areas underwent relatively extensive and successful Disarmament, Demobilisation, and Reintegration (DDR) processes, large numbers of ex-combatants in parts of Ituri District, Maniema Province, and northern Katanga Province now work as artisanal miners. For example, it is estimated that up to ten per cent of artisanal gold diggers in select parts of Ituri District are ex-combatants.[5]

The transition from war to peace in the DRC is, and continues to be, slow, with a completed phase of DDR and humanitarian aid and what should naturally evolve to a more strategic development and social recovery agenda. Within this framework of post-conflict transition, the security of communities remains fragile at best, as the pillars of local governance, economic opportunity and social cohesion rebuild at an alarmingly slow pace, bedeviled by individual opportunism and a culture of impunity. Insecurity is largely economic in nature, as the majority is unable to find sustainable livelihoods and thus continue to find recourse in artisanal mining.

Artisanal Mining at a Glance

The profile of artisanal miners in the DRC varies from province to province. One encounters individuals who abandoned school at the age of nine and now at 41 continue to work as gold diggers in Orientale Province. One can also encounter law students from the University of Lubumbashi panning for surface ore during

---

5   Based on Pact baseline research into demobilised fighters on the Ashanti Goldfields Kilo concession, July 2008.

school term breaks in the Kulu River of Kolwezi. The irony of artisanal mining in the DRC is that it exemplifies the greatest examples of Congolese ingenuity and entrepreneurship, and yet similarly reveals the greatest horrors of abuse, violence and exploitation.

As stated in the introduction, it is estimated that up to 2 million people work as artisanal miners across the country (WB 2008). With their dependents, it is estimated that the ASM sector provides economically for 18 per cent of the national population. Diamonds, copper, cobalt, gold, uranium and tantalum are the main resources extracted.

It is very important to view this statistic in global terms. Globally, large mines generate more than 95 per cent of the world's total mineral production. The industry employs an estimated 2.5 million people worldwide and is dominated by some 50 major mining and metals companies. Globally, ASM generates about 15 per cent of the world's minerals yet is a major source of income in about 30 countries around the world for an estimated 13 million people. Between 80 million and 100 million people are estimated to depend on small-scale mining for their livelihood with 52 million of them residing on the African continent.

The DRC is the reverse of this global situation. Approximately 90 per cent of minerals are produced by ASM and only 10 per cent by large-scale mining (Hayes 2008). This comparison serves to demonstrate the enormity of the transformation and the length of time that will be required for a peaceful formalisation of the mining sector.

Artisanal workers come from a variety of socio-economic backgrounds such as public security forces (current, demobilised and deserters), displaced farmers, and even skilled professionals. In the last decade, these individuals have become highly migratory, adapting to shifts in product demand across the country. Their temporary settlement in traditional communities can often have very negative social impacts including family break-up and polygamy, an increase in prostitution, abuse of alcohol and drugs, competition for—and destruction of—the communities' resources, and distortion of local market prices of basic goods due to their relatively higher daily income earned. Artisanal miners are frequently trapped in cycles of debt and poverty as a result of financial obligations to middlemen and women, known as 'negociants'.

In practice, artisanal miners' relationships with both traditional and local authorities are often governed by predatory taxation and bribery and their entry into indigenous sedentary villages is often viewed with mixed feelings by the local inhabitants. On the one hand, artisanal miners inject significant cash flow and attract services (such as kiosks, bars, and second-hand clothing stalls) otherwise unavailable in rural villages, while on the other hand, artisanal miners flout and attempt to coerce traditional and local government authority and impose a generalised social insecurity.

Vulnerability and Gender Issues: Fragile State Context

Pact's vision for its work in the DRC rests on re-establishing the security of communities wherein men and women play equally important, though perhaps different, roles in development. Security is not limited to physical protection but includes all aspects of life, most notably economic opportunity, health and wellbeing and community governance as expressed through social norms and traditional structures.

In any post-conflict fragile state, the period following an established cease-fire and peace process is critical as tangible 'peace' dividends must be felt by its citizens: people who have suffered prolonged and protracted conflicts tend to gauge the value of a peace period by what social, economic and political benefits they derive from it. Although this is often judged by basic social service delivery (such as education, health, water and roads), it also implies the security of individuals and their ability to perform daily livelihood functions without experiencing serious physical and/or psychological threats. For women, former combatants, and marginalised youth, it also implies the ability to actively participate in the economic and decision-making processes of their communities.

In the DRC, the current political, economic and social transition from war to peace continues to leave many of the root causes and drivers of the conflict unresolved. Within this framework of post-conflict transition, the security of communities remains fragile as the pillars of local governance, economic opportunity, and social cohesion re-build at an alarmingly slow pace. Within this void flourishes individual survivalism, impunity, and escapism—primarily by men—into alcohol and drugs. The vulnerability of women thus remains extremely high. In effect, women artisanal miners can be categorised as doubly 'at risk': they are rural women emerging from a war context and are additionally illegal workers living in precarious social, economic and environmental conditions.

This last aspect is important to consider. Societies learn to adapt during times of conflict with men often out fighting in the war and women home providing for families. Today one witnesses this continued adaptation in the migratory patterns of artisanal miners who leave wives and children in search of economic opportunities across the country. Rarely does money reach the family and women are in turn forced to seek livelihoods of their own. In concentrated artisanal environments, women will become labourers themselves or turn to supply businesses around the sites, including the sex trade. The vicious cycle of abandonment and resulting destruction of 'normal society' is witnessed when young females from as far as Kasai Province or North Katanga can be found working as prostitutes in artisanal areas along the southern border with Zambia.

Endemic to war and gender issues is also violence against women, which was used as an institutionalised tool of warfare in the eastern provinces of the DRC during the war. In artisanal mining communities, where high concentrations of ex-combatants reside, sexual and gender-based violence (SGBV) continues to be common practice.

As often with societies emerging from war and conflict, the traditional structures that once governed culture and social practice are either eroded or significantly diminished. Though Congolese women today would argue that the majority of past culture and social practice did not in fact promote respect and equal opportunity for women, they would agree that at least before the conflict, traditional structures existed for positive influence and change. Compounded with the usurpation of traditional authority and general disorder in ASM communities, gender re-definition becomes extremely complex as societies naturally evolve by grounding constructed roles and responsibilities in local leadership structures. The multi-ethnic composition of most ASM communities further complicates approaches to gender issues, as tribal values and practices vary significantly, and occur within the void of an overarching traditional governance structure.

For a society recovering from severe conflict, in order to restore reciprocity between genders as the means to 'normalising' social norms and hence community security, two complimentary avenues of approach are suggested. Firstly, the approach should address those ideas about what constitutes a man and, secondly, redress the powerlessness of women. In the first instance, and depending on the particular society's beliefs, attributes of maleness other than raw strength and power need to be brought to the fore. These positive attributes might include protection, fatherhood, responsibility and support. Secondly, women need to be empowered, to be recognised (not only by men, but by themselves) for the essential roles and attributes they bring in relation to men. Both of these avenues of approach need to be grounded in the material, economic parameters of their lives. In other words, men need to be reminded of other aspects of their manhood and given the opportunity to express these through work, the provision of economic support and protection of their partners. Equally so, women need to demonstrate the essentiality and importance of their contributions through work, the provision of economic support, and participation in decision-making.

The Development Entry Points

Within the above theoretical framework, Pact has chosen the following entry points for addressing gender mainstreaming in ASM.

Women-focused interventions that include:

- Literacy and savings training.
- Alternative livelihoods to transition out of artisanal mining.
- Female social capital for advocacy and decision-making in communities.
- Sexual and gender-based violence.

Men-focused interventions that include:

- Literacy and savings training for men to emerge from cycles of debt.
- Awareness on domestic violence.

Community interventions to improve artisanal work conditions, household and community economic earnings, and individual wellbeing, including:

- Workers' rights regarding child care and health conditions.
- Transition economies out of artisanal mining into agriculture and small business development.
- Primary health care and HIV/AIDS in artisanal mining communities.
- Community mechanisms to monitor and address sexual and gender-based violence.

Such an approach works with both men and women, and the community at large. These strategies reduce potential backlash on women, position the gender argument within the wider community and look towards long-term economic and social solutions.

Challenge for ASM within the Broader Development Agenda

ASM is provided for under the current 2002 DRC Mining Code. However, government regulated artisanal zones are extremely limited, forcing the majority of artisanal mining to occur in a void of government oversight and support. This makes interventions by interested partners limited. International agencies and donors are hesitant in general to be seen working closely with the extractive industry sector in the DRC and even more so with ASM given its associated reputation with corruption, bribery, and abuse.

For example, at the time of writing, in Katanga Province only six development projects from private and public sources target ASM issues. Of these programs, none have a specific gender focus though women are beneficiaries of child education, health awareness, and economic alternatives activities. In Ituri District, there are only two projects that currently focus on ASM. Neither targets women directly.

Contributions by industrial companies are equally limited as companies shy away from intervening in a sector that is still largely illegal and exploitative. The most common strategy used is the offer of alternative employment. However, artisanal women, who are largely uneducated, rarely fulfill employment criteria for industrial recruitment as it often entails high-intensity manual labour or operating of machinery. As confessed most recently by a senior mine manager, affirmative action to recruit women is not easy as 'they just don't have the base qualifications'. Thus women rarely gain from transition opportunities due to their inability to physically respond to manual labour positions or due to their lower education background.

Justification for 'Women in ASM' Development Focus

If limited funds and interest exist for ASM sector reform in the DRC, what then becomes the case for a focus on women's interests in particular?

Firstly, women may constitute upwards of 50 per cent of the current ASM labour sector and carry greater economic burden and responsibility than their male counterparts. In some mining areas, women on average constitute up to 70 per cent of single ASM households with an average of six child dependents. This includes widowed, abandoned and divorced women as well as those with unemployed husbands. Yet as already outlined, they are often the least considered in ASM programming in the DRC.

Secondly, working with women artisanal miners often increases sustainability of economic livelihood alternatives and enhances social change as women are not as mobile as men when seeking ASM opportunities, and thus they generally remain rooted in their communities. They constitute a bridge to more traditional social life and structures, and thus increase the chances of influencing the broader dynamics for behaviour change.

Thirdly, in the experience of Pact staff and researchers, women are more likely to abandon mining and work on economic transition opportunities if given the chance. They are less reticent towards alternatives and do not make high demands for compensation. Though men in general have proven to be far more entrepreneurial when it comes to taking new business risks, female artisanal miners remain committed to an opportunity and save more from their earnings over time.

However, as with any other attempt to address equity and equalisation of gender roles, programming focused solely on women can quickly alienate them from the broader society and make men even more resistant towards change. Especially concerning sexual violence against women, the roots are often found in male feelings of powerlessness and social marginalisation. By focusing on women alone, efforts could in fact produce violent backlashes and make women more insecure than before. Thus any efforts to address gender issues in the ASM sector need to be within a broader framework for change where all parties find their interests represented.

Women and ASM: Katanga and Orientale Provinces

The following discussion touches upon Pact's experience with partners across two distinct geographic areas of the DRC. At the time of writing, Katanga Province is undergoing an industrial mining renaissance, with companies mining primarily copper and cobalt. In the northeast of the country, Ituri District of Orientale Province is slowly recovering from the impact of conflict with a few major companies considering operations, mainly in the gold sector.

Economic Empowerment and Livelihood Alternatives

Women work in and around artisanal mines, most often as processors and transporters of raw materials. They also act as service providers to the mine in the areas of commerce, catering and, frequently, prostitution. The peripheral nature of these roles has significant impacts on the potential livelihoods for women, as they rarely participate in the core mining activity and thus do not have a voice in operational decision-making. In Katanga Province, examples do exist of women 'negociants', the middle person who sells raw ore to traders, but such examples are rare.

In 2007, Pact with support from the International Finance Corporation (IFC), USAID, and four major mining companies (Tenke Fungurume Mining, Anvil Mining Ltd, Katanga Mining and DCP/Nikanor) completed an eight-month research project into the artisanal mining sector of Kolwezi, Katanga Province. A socio-economic survey interviewed 255 artisanal women miners to ascertain levels of income, social circumstances, and barriers faced in the ASM sector.

The survey revealed that single, divorced or widowed women make up 36 per cent of the female population, having to care for six children on average. Thus, this group can be effectively classed as female-headed households. Such a large percentage of female-headed households in any given population is an indication of several underlying conditions. Firstly, it indicates by virtue of the absence of male heads that nuclear family norms have been disrupted, either through such avenues as poverty and loss of assets or through migration of males or through generalised breakdown of overall societal norms.[6] Secondly, it indicates a greater dependency ratio of children to adults, with single women carrying a greater burden to provide for their children than standard nuclear joint-spouse households. Thirdly, it indicates diminished access to resources in these households, due to limited household assets such as land (normatively accessed through males) and unequal access to employment and fair wages by virtue of being women and subject to economic discrimination.

The number of these vulnerable female-headed households is augmented by a further 52 per cent of married women's households in which the spouses are unemployed and thus dependent on a single source of income. In effect, 70 per cent of all of the women surveyed were the sole income sources for their families.

In response to such staggering figures, Pact with Anvil Mining Ltd has been modeling economic transition alternatives with female artisanal miners in Kolwezi. The program combines literacy and savings programs with vocational training and technical accompaniment. Over the course of one year, 80 former artisanal women miners have successfully transitioned out of artisanal mining into other economic opportunities. In most cases, women are pursuing two income-generating activities due to the seasonal nature of agriculture, thus combining farming with small businesses in town such as bakeries and restaurants. Beyond the tangible economic and health benefits, women attest to a greater sense of self-worth and confidence provided by the literacy program.

Economic transition activities do present challenges, though easily overcome if recognised at the outset. Of greatest importance is ensuring income is available during the initial transition period as artisanal miners are used to earning income on a daily basis. When presented with business opportunities where initial income will not be generated, creative solutions to bridge the earning gap must be identified and supported.

---

6   There is one exception to this generalisation and that is when female heads are dependent on remittances from spouses who have migrated successfully in search of employment—this exception does not appear to be the case for Kolwezi given that it is itself is a magnet for labour migration.

Reproductive Health

Health and safety is major area of concern in the ASM sector as a whole. Men and women alike suffer from a variety of respiratory illnesses, many chronic due to long-term exposure to highly mineralised ore bodies, or mercury used in gold recovery processing.

For women, the impact of these working conditions has generational consequences. In Kolwezi, alarming rates of stillbirths, miscarriages, and birth of deformed babies are documented by the Small-Scale Mining Technical Assistance and Training Service (SAESSCAM), the government body responsible, and Paraclisis Research Group at the University of Lubumbashi. These cases are most common with women exposed continuously to highly radioactive substances such as uranium, copper and cobalt. Out of 350 children surveyed in 2007 by Paraclisis Research suffering from respiratory problems, 41 per cent had parents who worked as artisanal miners and frequently accompanied their parents to the mine site. The ignorance regarding the health consequences of artisanal mining is high and necessitates concerted efforts in education awareness campaigns.

Further, across artisanal sites and camps, women have little exposure to health prevention measures. Women sex workers are frequently forced to engage in unprotected sex, increasing their chances of contracting sexually-transmitted diseases (STDs), and HIV/AIDS. In partnership with USAID, Pact has trained 20 women artisanal miners in basic health and reproductive education in Kawama, the Katanga Province's first official artisanal zone. This sensitisation created a demand from women in the market to sell condoms at their kiosks. Pact couples their rural women's health outreach programs with artisanal women working and living in adjacent communities.

Sexual and Gender-Based Violence

Perhaps the most troubling aspect of women working in the artisanal mines and camps is their exposure to a violent and volatile community dominated by men unattached to either family, traditional community or place. During his March 2008 DRC visit, the former UN Special Envoy for Aids in Africa, Stephen Lewis, remarked that, 'the DRC is by far the worst place in the world for women. The destruction of women is beyond the capacity of the mind to absorb.' Though most of the media and advocacy efforts focus on the problem of sexual and gender-based violence (SGBV) in eastern Congo, two recent surveys carried out by Pact reveal alarmingly high rates of SGBV in and around artisanal mining environments in Katanga Province and Ituri District.

The research showed that reported SGBV fell into four categories: sexual assault and rape, prostitution, forced marriage (*le mariage précoce*) and domestic violence. Sexual assault and rape was further assigned to three kinds:

1. Predations against girls and women of all ages by individuals or groups of men, usually under the influence of alcohol or drugs and including police and military personnel.

2. Violent assault and rape of girls under the age of 11, mostly victims of local sorcery prescriptions to individual men for the acquisition of wealth and virility.

3. Gang rape of girls between the ages of 12 and older participating in parties, street revelries or other community social festivities.

This research corroborates initial data found in a UNICEF qualitative study[7] on GBV against children in the artisanal mines of Katanga Province.

The presence of SGBV in such socially unbalanced circumstances as mining camps should not be surprising. Given the weakening of traditional village authority together with the overwhelming predominance of single men, coupled with the already inferior social position of local women, especially the younger women and girls, alongside the more worldly women providing services to the camps, the unmitigated vulnerability of women and children is exacerbated.

Sorcery, fetishes and superstitions also heavily influence violence against women in all social contexts of the DRC. In the ASM sector, sex with young virgins, often children as young as five, is encouraged by witch doctors as a means to secure wealth. In one month alone in 2008, Pact received reports of the rape of three children; two of which were between the ages of two and four years old. In one case, the child was left in the woods after having been raped and the local chief discovered a group of men attempting to burn her alive suspecting her to be a witch.

In select parts of Ituri District, increasing rates of SGBV are linked to artisanal mining, and according to local partners working in one highly dense artisanal mining town, Mongbwalu, male miners constitute the main perpetrators of SGBV incidences.[8] In 2007 alone, 1289 SGBV cases were reported in Mongbwalu and its surrounding area, with less than 20 per cent addressed. This is in comparison to a total of 1881 cases reported for the entire Ituri District in 2007 by the UN coordination mechanism.[9]

---

7   The author attended the de-briefing at the UNICEF office in Lubumbashi where the consultants hired by UNICEF presented their findings. At the time of writing, the UNICEF report had not been released.

8   The Mongbwalu Stakeholder Forum has kept records since 2007.

9   Source: UN Office for the Coordination of Humanitarian Affairs SGBV Coordination, Bunia. January 2008.

In Katanga Province, the vulnerability of women and children living within ASM communities is equally staggering. Here, as is common elsewhere, incidents of SGBV are consistently underreported, so that the figures in Table 10-1 are only indicative of a much larger problem. In fact, local professionals working in the sector estimate under-reporting as great as 80 per cent.

Table 10-1: Reported incidents of rape in Katanga Province, Jan–Mar 2008.

January 2008	February 2008	March 2008	Total
40	141	153	334
Only 4% of health centres reporting	Only 25% of health centres reporting	Only 30% of health centres reporting	28 centres reporting out of 67 total health centres

Note: under-repoting error exists for all data. Source: UNFPA 2008.

However, male artisanal miners are not the sole perpetrators of SGBV in and around ASM areas. The presence of police and army officials on industrial mining concessions, and in ASM zones, also contributes to increased insecurity for women. In most cases, a family will settle a rape case with a public security officer for as little as a goat.[10] Rarely are cases reported to the authorities and even more rarely do they make it through the judicial system.

General prevention and reduction of SGBV incidents requires a varied approach including treatment of victims, strengthening of judicial systems, redefinition of gender roles in communities, and economic and social re-integration for victims and their families. With mining companies and UN partners, Pact is piloting models of prevention and reduction of SGBV in both Katanga Province and Ituri District. These models build on current UN coordination mechanisms for reporting and addressing SGBV while introducing stronger economic and gender re-definition aspects. This includes literacy and savings programs to build positive male social capital, reduce indebtedness towards 'negociants' and traders, and allow for economic transition opportunities to facilitate family reunification and normalisation of social relations.

Also, the presence of social development projects in and around heavily-populated ASM sites is slowly contributing to behaviour change, particularly around reporting of incidences. This is important as a first step to reducing SGBV, as silence and fear of victims and their families to speak out against these violations encourages on-going impunity of aggressors. Through Pact's women's literacy and savings program, WORTH, women are reporting incidents of rape and sexual violence more frequently. Through a monthly human rights and security meeting between the security heads of mining companies, private security companies, public security forces, the UN, and Pact, incidents of rape and intimidation are reported on concessions. In addition, mining companies

---

10   The average market price of a goat is between US$50–60.

have committed to train their public and private security agents in SGBV related laws and penalties in the DRC and establish SGBV monitoring and reporting mechanisms on their concessions.

Children and Education

A troubling impact of ASM practice relates to the children of women workers. A plethora of factors drive women to bring their children to the mine site on a daily basis, including the need to supplement family income, lack of affordable education facilities, and insufficient child care alternatives. It is not uncommon to find babies sleeping under trees in the shade, with older children assisting in mining activities such as sorting and washing. In Kolwezi, it is estimated that 23.7 per cent of child miners work with their mothers at site (Pact Inc. 2007).

Women cite school fees as a main component of household income supplemented by their ASM work. Free primary education is not universally provided for in the DRC and even in areas where it is, it is often over-subscribed and under-resourced. Through a partnership with Solidarity Centre and Save the Children UK, and with funds by the US Department of Labor and Anvil Mining Ltd, Pact is working to improve education facilities and resources, staff training, and after-school activities in ASM-vulnerable communities in Kolwezi and Mongbwalu.

Leadership and Representation

The DRC's regulatory environment and capacity are weak, and existing laws such as the Mining Code are not enforced effectively. A range of government actors perform specific roles and responsibilities within the ASM sector, though much confusion exists around actual field practice. The manipulation of these mandates is a continued source of frustration for artisanal miners that drives suspicion, discrimination and exploitation, and often leads to violent conflict at mine sites.

The first key government actor is the Small-scale Mining Technical Assistance and Training Service (SAESSCAM) that theoretically provides technical input and support to artisanal miners, including protection and safety. The second is the Division des Mines who provide for an inspector at the mine sites. The third are the authorities tasked with creating artisanal mining zones and issuing cards for artisanal mining, trade and transport. Finally, the state is a buyer, through government bodies such as Gécamines (Katanga Province), MIBA (Kasai Orientale Province) and OKIMO (Orientale Province). However, given the lack of specific artisanal zones established and functioning in most parts of the country, the majority of these government bodies is entangled in webs of predation and exploitation, and is often highly distrusted by the very people they are meant to protect and represent.

Cooperatives should logically emerge to fill the gap in veritable representation of interests. To date in the DRC, very few cooperative or association structures do in reality perform such a role. Rather most are synonymous with high 'membership fees', that is taxes and few benefits. At best, these structures pay for ad hoc requests such as hospital bills and funerals. None of the above structures have specific gender policies. It is one of the most serious gaps the ASM sector faces for ensuring effective representation of women's issues.

Moving Forward

Artisanal mining will remain an important economic recourse in the coming years for thousands of individuals across the DRC. In fact, at the time of writing, it is anticipated that ASM may even increase as a result of several factors, including:

- liberalisation of the ASM sector in certain Provinces by government, such as in Orientale Province;
- continued instability in eastern Congo that fuels access to resources, and forces displaced persons to seek economic livelihoods outside their home areas;
- limited economic alternatives following industrial consolidation of mining concessions;
- on-going high demand for mineral resources by countries such as China, India and Russia; and
- a current global economic recession that has halted the majority of large-scale mining projects in the DRC.

However, as this chapter demonstrates, the challenges facing the sector are great, let alone those that are gender specific. How then to build on the few lessons that do exist?

Overcoming Challenges: Lessons and Achievements

As previously stated, Pact has been working with a range of development actors, foreign donors, mining companies, and government to respond to the challenge of gender issues in the ASM sector of the DRC. This has formed part of broader efforts to increase understanding of ASM as an integral part of the broader development agenda for the DRC. The result has been increased collaboration between stakeholders, leading to a range of achievements and lessons learnt, including:

- *Securing a policy framework for coordinated action and investment*: SAESSCAM, with technical support from Pact, has proposed an ASM framework for the DRC to encompass all potential partners and funding sources over the coming five years. Gender mainstreaming is an integral part of the framework.

- *Lobbying the Government and UN agencies to respond to specific women's sectoral challenges that cut across ASM and rural populations*: HIV/AIDS and SGBV are two examples where Pact places great emphasis on integrating women artisanal miners into regular development programming.

- *Establishing public-private partnerships that respond to both global development objectives and standards, and that are in line with the national development plan of the DRC Government*: Whereas companies may not wish to fund ASM reform, public funds to support initiatives within a broader extractive industries public-private partnership is possible and has proven to work. Through USAID funding, four industrial companies in the DRC have benefited from select ASM initiatives by Pact and partners.

- *Creating small pilot opportunities for collaboration with UN agencies and other development organisations*: The Katanga Provincial Ministry of Mines is workorking in collaboration with UN Agencies and Pact to establish its first artisanal zone in Kawama. To date, a joint UN-Pact mission established baseline needs and statistics, and ongoing technical support is provided to the Ministry. A strong component of this support is to improve access to health and reproductive services for women artisanal miners, including education and testing facilities through local clinics.

- *Integrating artisanal populations into community development projects if in the same operational area as a publicly-funded development program*: Pact includes artisanal women into their literacy and savings' programs for rural women in addition to their education work with women on HIV/AIDS.

Appendix 10-1: Pact's Gender ASM Partners in Katanga Province and Ituri District in Orientale Province.

ASM partner	Sector	Location
DRC Government		
Ministry of Mines	Artisanal zones	Katanga Province
SAESSCAM	National framework for ASM, economic transition	Kinshasa and Katanga Province
Donors		
USAID	SGBV, economic transition, HIV/AIDS	Katanga Province and Ituri District
US Department of Labor	Child miners education	Katanga Province and Ituri District
IFC	Research into ASM	Katanga Province
UN agencies		
ILO	Labor practice, representation and leadership	Katanga Province
UNFPA	SGBV, HIV/AIDS, STDS, Health and reproduction	Katanga Province and Ituri District
WHO	HIV/AIDS, STDS, Health and reproduction	Katanga Province
Unicef	SGBV against child artisanal miners	Katanga Province
UNDP	Economic transition	Katanga Province
International NGOs		
Pact	SGBV, economic transition, leadership, WORTH, and HIV/AIDS	Katanga Province and Ituri District
Save the Children UK	Prevention of child miners through education and support to artisanal mothers	Katanga Province and Ituri District
Solidarity Centre	Prevention of child miners through education and support to artisanal mothers	Katanga Province and Ituri District
Groupe One	Prevention of child miners through education and support to artisanal mothers	Katanga Province
National NGOs, universities, local associations and networks		
University of Lubumbashi	Health and safety of women	Katanga Province
Réseau Action Femme	SGBV	Katanga Province
WORTH	SGBV, savings and microcredit and small business	Katanga Province and Ituri District
Mining Companies		
Anvil Mining Ltd	SGBV and economic transition	Katanga Province
Anglo Gold Ashanti	SGBV, economic transition and employment	Ituri District

References

HHI (Harvard Humanitarian Initiative), 2009. 'Characterizing Sexual Violence in the Democratic Republic of the Congo. Profiles of Violence, Community Responses, and Implications for the Protection of Women.' Final Report for the Open Society Institute. Viewed 20 January 2011 at http://globalsolutions. org/files/public/documents/GBV_final-report-for-the-open-society-institute-1.pdf

Hayes, K., 2008. 'The Economics of Artisanal Mining in Africa.' Unpublished Pact Inc. Report. Viewed 9 September 2010 at http://empoweringcommunities. anu.edu.au/documents/Perks%20paper%20&%20cover%20page.pdf

Hayes, K. and R. Perks, 2011. 'Women in the Artisanal and Small-Scale Mining Sector of the Democratic Republic of the Congo.' In S.A. Rustad and P. Lujala (eds), *High-Value Natural Resources and Post-Conflict Peacebuilding*. Earth Scan Publications.

Hinton, J., M. Veiga and C. Beinhoff, 2003. 'Women and Artisanal Mining: Gender Roles and the Road Ahead.' In G. Hilson (ed*.), The Socio-Economic Impacts of Artisanal and Small-Scale Mining in Developing Countries*. Netherlands: Swets Publishers.

ICRC (International Committee of the Red Cross), 2004. 'ICRC Annual Report 2004: Democratic Republic of the Congo.' Viewed 9 September 2010 at http:// www.unhcr.org/refworld/docid/46938ffc0.html

Pact Inc., 2007. 'Economic Development and Governance Transition Strategy for Kolwezi, Katanga Province.' DRC: Pact Inc.

UNFPA, 2008. '*Rapport des Nouveaux Cas des Violences Sexuelles Survenus au Katanga au Cours du Premier Trimestre, Mars 2008.' Initiative Conjointe de lutte contre les violences sexuelles faites aux femmes, aux jeunes, aus enfants et aux homes en RDC.* [Report of new cases of sexual violence surveyed in Katanga during the First Quarter, March 2008, UNFPA: Joint Initiative to combat sexual violence against women, young people, children and domestic violence in the DRC.]

World Bank, 2008. 'Democratic Republic of Congo Growth with Governance in the Mining Sector.' Report No. 43402-ZR, Oil/Gas, Mining and Chemicals Department, AFCC2, Africa Region. Washington: The World Bank.

WHO (World Health Organization), 2006. *Working Together for Health: The World Health Report 2006*. Geneva: World Health Organization.

11. Artisanal and Small-Scale Mining: Gender and Sustainable Livelihoods in Mongolia

Bolormaa Purevjav

Introduction

The practice of artisanal and small-scale mining (ASM) has grown considerably in Mongolia since the country transitioned from a socialist state to a market-based economy with a heavy reliance on mining and resource extraction. Calls to ban ASM due to lack of regulation and environmental and health concerns have failed to gain support because of the large numbers of the rural poor that have adopted ASM as a livelihood strategy to supplement diminishing incomes from agriculture. Instead, the Mongolian government has sought to regulate and reform the sector by offering skills training and increasing the capacity of ASM collectives. Women are actively involved in ASM in large numbers in Mongolia, as they are in many other parts of the world, due to their exclusion from employment in large-scale mining and deepening rural poverty. Despite policy commitments to gender equality, women in Mongolia struggle to find equal employment and women are attracted to ASM due to the absence of barriers that characterise employment in the formal sector. Mongolia's adoption of the Millenium Development Goal 3 (MDG 3), namely to 'promote gender equality and empower women', has resulted in the acknowledgement and incorporation of gender in the ASM reform program. This chapter describes the gendered practice of ASM in Mongolia and how the government in partnership with the Swiss Agency for Development and Cooperation (SDC) is attempting to reform the ASM sector in a gender-sensitive way through the Sustainable Artisanal Mining Project.

Mining and Economic Development

Mongolia is one of the least populated countries in the world. The territory of Mongolia measures 1 564 116 square kilometers—the nineteenth largest country in the world—yet has a population of only 2.6 million people. It is a landlocked country between Russia and China, located on the plateau of Central Asia; far from sea ports and transport networks. Mongolia's geography features high mountains in the west, wide open steppes in the east, the Great Gobi Desert in the south and the Taiga Forest in the north. Its climate is continental; long winters with frequent snow fall, springs with persistent winds, hot and very dry summers, and a short agricultural season. In particular, this geography and climate significantly affects the living conditions of rural people in Mongolia and presents a challenge to livestock husbandry and agriculture. Unaffected by these harsh climactic conditions, Mongolia's mineral wealth has become a key part of its economic development in the past two decades.

From 1921 to 1990 Mongolia was a socialist country. The socialist state system provided universal access to social services and all people were employed in state-owned cooperatives and institutions. In general, it offered considerable security and a decent standard of living. However, the socialist system ignored the desires and aspirations of individuals, curbed people's political freedom and was based on a centralised, authoritarian control.

In the early 1990s, when the socialist system became incapable of governing the country effectively, radical political and economic changes were introduced to re-define the country's economic development. Since then, Mongolia has undergone dramatic changes: foreign relations have expanded; rural-urban socio-economic disparities have deepened, significantly contributing to internal and international migration; a new constitution was approved and the legislative, executive and judicial powers are now distributed amongst the presidency, parliament, government and the supreme court. Today Mongolia is a democratic country that respects human rights and freedoms and has adopted a market-based economy, based on foreign investment, the expansion of the private sector and multiple forms of individual property ownership.

During the initial years of transition from a socialist to a market-based economy (from 1990–1993), Mongolia's gross domestic product (GDP) decreased by 20 per cent and many industries collapsed. The economic downturn resulted in high unemployment and a dramatic increase in poverty. From 1994–2002 Mongolia experienced rapid economic development and GDP grew by two per cent per capita on average. In 2003 GDP grew by 5.6 per cent, in 2004 by 10.6 per cent, in 2005 by 7.1 and in 2006 by 8.4 per cent. The acceleration of economic growth was largely based on the expansion of the agricultural, mining and service sectors. The formal, large-scale mining sector has been the largest contributor to Mongolia's economic growth—growing around ten per cent every year since 1991. In 2007, the mining and extractive industries constituted 27.5 per cent of GDP, agriculture, hunting and forestry constituted 20.6 per cent, wholesale and retail trade 14.2 per cent and the transportation and storage sector 10 per cent (NSOM 2007).

Unfortunately Mongolia's economic growth has not benefitted all segments of society. Despite consistent GDP growth over the past two decades, few new jobs have been created and poverty and unemployment remain high. According to recent research, 36 per cent of the population still lives in poverty. Rural areas remain the poorest due to lack of employment opportunities. People living in rural areas have responded to the challenges brought about by the structural adjustment of the economy by working in the informal sector—in particular, artisanal and small-scale mining in increasing numbers (UNDP Mongolia 2008).

Development and Growth of ASM

Artisanal and small-scale mining (ASM) developed in Mongolia in the early 1990s in response to several factors; diminishing employment and livelihood opportunities in rural areas being the primary cause. Mongolia experienced a series of particularly severe winters with heavy snowfall between 1997 and 2002, which exacerbated levels of poverty and hardship for those dependent on agriculture and animal husbandry. An alarming rate of depletion of livestock through uncontrolled use of natural resources by powerful elites has also threatened rural livelihoods in recent years. These factors, combined with the relative geological wealth of Mongolia's rural areas and the introduction of the Liberal Mining Law in 1997, which opened up 45 per cent of the country to mineral exploration (including areas that had been used for herding and breeding livestock) and greatly expanded foreign-owned mining activity, saw many rural people turn to ASM as an alternative livelihood (see Table 11-1).

Table 11-1: Number of artisanal miners: 2000–2008.

Year	Number of artisanal miners
2000	20 000
2003	100 000
2006	96 000
2007	67 000
2008	54 000

Note: the decrease in artisanal miners in 2007 and 2008 was due to confiscation of ASM mills by the government and prohibition of milling operations using mercury. Source: MRPAM (2007).

ASM refers to mining by individuals, groups and families. It is an activity of simplified mineral extraction of primary and secondary deposits. ASM activities are typically highly mobile and labour-intensive—the generally small deposits are usually exhausted after only a few years and tools and extraction methods are rudimentary and manually-operated. In contrast to large-scale mining, ASM has only two phases: extraction and processing. Most significantly however, is that ASM creates far more jobs in rural areas than large-scale mining and thus holds greater potential for reducing poverty.

ASM has developed and expanded informally in Mongolia over the past 15 years, until February 2008, when the Government of Mongolia approved the temporary regulation of ASM. This temporary regulation has provided artisanal miners with rights to mine but with limited land rights. Prior to this temporary regulation, ASM was regarded as highly disorganised and environmentally damaging by the central government, as well as large mining companies and environmental agencies, who lobbied to prohibit it.

As with other informal livelihood activities, ASM presents numerous human and environmental hazards: health problems due to harsh weather conditions and poor occupational safety; limited access to social services and assistance; lack of training in the knowledge and skills required to undertake mining activities; conflict with others; illegal digging in the licensed areas of large mining companies and pasture degradation. Unsecured mine sites pose dangers to children and animals, unsafe use of toxic chemicals such as mercury has led to disease and ASM has increased prostitution and the spread of sexually transmitted diseases (STDs) and HIV/AIDS, as well as the use of child labour.

While all of these hazards present regulation and legal challenges for governments, in Mongolia artisanal miners currently comprise 20 per cent of the employed or 'economically active' population—three times the number employed in the large-scale mining sector (ADB Mongolia 2008). Due to the high number of people engaged in ASM and the relatively good income levels it generates, the government had little choice but to take a positive step towards recognition of ASM as a source of 'formal' employment by issuing the temporary regulation.

ASM and Sustainable Livelihoods

Unofficial statistics suggest that ASM currently provides employment for approximately 100 000 people and produces an estimated 5–10 tonnes of gold per annum, generating an estimated annual production revenue of between US$60–120 million. ASM is practiced at over 100 sites across 19 provinces in Mongolia. The majority of artisanal miners engage in hard and placer gold mining (80–90 per cent), with the rest engaged in fluorspar and coal mining. As these figures demonstrate, ASM provides important development opportunities from both a macro-economic and socio-economic perspective. ASM is critically important for local communities as it provides vital temporary and full time work, despite its negative health and environmental implications. As artisanal miners use simple tools and equipment which do not require specialised skills, the overall entry cost to ASM is relatively low. Thus it provides an opportunity for the very poor to earn an income.

Given its current rate of expansion and levels of participation, ASM operations have reached the benchmark of a sub-sector in Mongolia. Rather than ban ASM, the state provides support in terms of the establishment of a legal framework in which ASM can safely operate although organisational and institutional capacity building have been called for in order to protect livelihoods. Despite the temporary regulation of ASM by the Mongolian government, obstacles remain to its inclusion in the mainstream economy. It is still not possible to sell ASM-derived gold to the Bank of Mongolia and currently all gold produced by artisanal miners is smuggled by tradesmen over the green border into China. The economic potential of ASM is further constrained by poor work practices and standards of processing. The absence of a regulatory framework and secure rights for artisanal miners have exacerbated existing problems and conflicts both within the sector and between ASM and the formal mining sector.

In order for local communities to maximise the benefits of ASM, the government needs to integrate ASM into rural development strategies and support a range of reforms. These include establishing a regulatory framework and the introduction of workplace training and standards of practice, such as the safe use of toxic chemicals, disposal of toxic waste and occupational health and safety. To improve the extraction of deposits, promotion of more efficient ASM tools and processes is needed. This will also bring environmental benefits and help to make ASM a more sustainable industry. Lastly, a regulatory framework with a clear definition of the rights and responsibilities of artisanal miners would contribute to local development and ensure that artisanal miners can become valued members of society.

Because the challenge of sustainable development is to achieve a steady economic growth rate and address social needs while not depleting natural resources, the development of ASM requires a holistic approach that considers the environmental, economic and social challenges together (see Figure 11-1). The development of an ASM policy framework for Mongolia also requires strong partnerships and cooperation between all relevant stakeholders, including international organisations, to tackle the complex issues involved and enable ASM to act as a 'motor' for sustainable rural development.

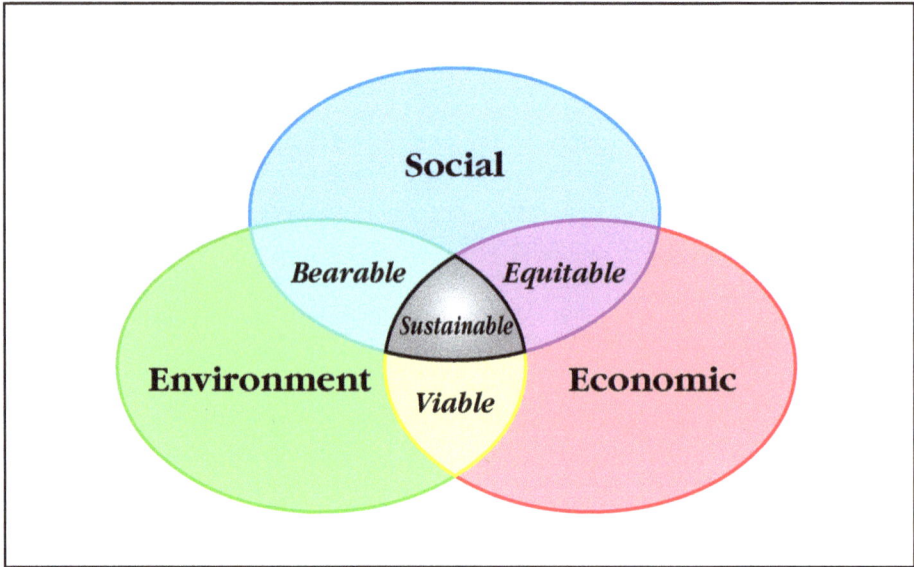

Figure 11-1: The heart of sustainable development: addressing economic, environmental and social aspects equally.

At all levels of the Mongolian government there is a belief that large-scale mining is the best option for economic development. The argument is that because it is easier to control, it contributes tax revenue and therefore has stronger national economic benefits. However, international experience shows that ASM can be a motor for rural development. Examining ASM purely from a social perspective, we find it creates much-needed work and is a major source of income in rural areas. Although there are environmental impacts and risks, artisanal and small-scale miners typically mine the smallest deposits, and rework dumps and tailings. From a macroeconomic perspective, ASM offers further benefits in terms of multiplier effects: it mobilises natural resources, generates revenue, activates trade and contributes to local development through money circulation, investment activities and demand for product and services.

If effectively regulated, ASM could also contribute to the achievement of Mongolia's Millennium Development Goals (MDGs). Specifically it could contribute to achievement of the following goals:

- Reduce by half the proportion of people living on less then one dollar a day.
- Promote gender equality and empower women.
- Combat HIV/AIDS and other diseases.
- Ensure environmental sustainability.
- Develop global partnerships for development.

Despite these benefits, both real and potential, ASM entails numerous costs. From a geological perspective, the costs include exploitation of a nonrenewable resource; losses through unsustainable exploitation of the highest grade minerals; low recovery due to mining methods and the impacts of mineral transportation. The environmental costs include deterioration and destruction of surrounding land, soil, ground and surface water, air, flora and fauna, energy sources and ecosystems. In terms of social costs, ASM entails the aforementioned low occupational safety standards and high health risks as well as poor living conditions, complex economic dependencies, lack of social security, alcoholism, crime, prostitution, child labour, violation of local community rights and conflict (UNFPA 2007).

Gender and Development

The state of gender relations and equality in Mongolia offers a contradictory picture. On the one hand, Mongolia has relatively high Gender Development Indicator (GDI) figures, has embraced MDG 3 and has a number of progressive gender equity policies in place. Women's mortality rate has fallen in recent years (although rates in rural areas remain high) and women's level of educational attainment is often higher than that of men (ADB Mongolia 2008). Despite these gains, in practice women continue to receive less pay than men, have fewer employment opportunities and are found in less managerial positions across professions and sectors. Despite recent strong economic growth, inequalities are increasing, particularly between urban and rural areas and women in particular are entering the informal sector in greater numbers (including ASM), with fewer protections and lower wages. Gender relations in Mongolia are also being transformed by high rates of male suicide and early mortality, alcoholism, gender-based violence and unemployment (ibid.).

In terms of gender equality legislation, the Mongolian Constitution, Labor Law, Family Law, Civil Code and Criminal Code contain provisions to promote gender equality. The Constitution provides equal rights for women and men in political, economic, social, cultural and family life, and prohibits all kind of discrimination based on sex. The 1999 Labor Law also prohibits gender-based discrimination in employment and contains a section addressing maternity rights. The approval of a Domestic Violence Law in 2004 represented a significant step in strengthening women's rights and efforts to reduce violence against women. The amendment to the Criminal Code in 2008, which included a full definition of trafficking according to international standards, is another milestone in the area of gender equality and protection of women's rights.

In 2005 the Mongolian Parliament passed a resolution on MDG 3 in order to improve the social, professional and cultural status of Mongolian women and promote their participation in broader development policies and programs. The MDG-based long-term National Development Strategy (NDS) was adopted by Parliament in February 2008 (NDS 2008). The NDS is the main policy document that guides and coordinates all national policies, programs, and laws with sector strategies aligned to NDS mandates. The gender equality policy for NDS is to 'ensure human rights-based gender development through universal gender education and provision of gender equality in the labor market...', and contains provisions to increase the number of female candidates for parliamentary elections.

The key government mechanism for gender equality, the National Committee on Gender Equality (NCGE) was established in 2005 and is chaired by the Prime Minister. The NCGE has 33 members with representatives from government, the private sector, and NGO/civil society groups. The members are divided into five working groups to oversee thematic priorities, which include legislation; implementation of Committee on the Elimination of Discrimination against Women (CEDAW); gender research and monitoring; governance; and public relations. However, in practice most of the representatives have little knowledge of gender equality issues and lack a clear commitment to implement the NCGE.

Some of these legislative changes and policy measures are clearly having a positive affect on gender relations in Mongolia, although the statistics offer conflicting evidence. The Human Development Report of 2007 demonstrates that the Gender-Related Development Index (GDI) in Mongolia has risen in recent years, from a GDI of 0.693 in 2004 to 0.719 in 2006. However, the gender empowerment measure (GEM) worsened from 0.448 in 2002 to 0.422 in 2007, and is lower than its gender development index (GDI) value of 0.719 for 2007 (UNDP Mongolia 2008), evidencing inequality in economic and political opportunities.

Just as Mongolia's progressive gender policy framework obscures disparities in terms of income and labour market opportunities between men and women, Mongolia's increasing GDP has not decreased rates of poverty and inequality despite high literacy and school enrollment rates (see Table 11-2). A recent Country Gender Assessment provided evidence that the rapid economic growth in Mongolia over the last four to five years has not translated into a secure future for many men and women and that inequality based on income, gender and social exclusion is increasing. The poverty rate remains high, particularly in rural areas (37.9 per cent) compared to urban rates of poverty (27 per cent). Female-headed households continue to be poorer (43.8 per cent) compared to male-headed households (34.8 per cent). In rural areas, herder households are amongst the poorest, closely followed by female-headed households (ADB Mongolia 2008).

Table 11-2: Mongolian Gender Indicators: 2006.

Gender Indicators	Male	Female
Life expectancy index (%)	62.6	69.3
Adult literacy rate (%)	–	97.5
Combined primary, secondary and tertiary gross enrollment ratio (%)	75.5	83.2
GDP per capita (PPP)	3.1	2.6

Source: UNDP Mongolia (2008).

According to the Mongolian Office of National Statistics, in 2006 women made up over 50 per cent of the economically active population and 51.6 per cent of the employed workforce. Women represent a large proportion of clerks, service workers, professionals and technicians. The high female labor force participation vis-à-vis men reflects in part a reduction in the male working age population due to rising levels of male disability caused by working in high risk environments (chiefly the construction and mining sectors) where low occupational safety is in place, and high levels of mortality due to alcoholism and high health risk behavior.[1]

The high level of female workforce participation is not enough, however, to guarantee gender equality and the gender pay gap is widening. This is particularly evident in the education sector, where women make up the majority of employees but receive 87.6 per cent of average male salaries. Men in managerial positions across all sectors receive 22.8 per cent more in terms of salaries than women in similar positions. Women's pensions are also lower than men's and are a contributing factor to the relatively higher rates of female poverty.

The double work burden, care obligations and inadequate access to credit are the main disadvantages and obstacles women face in terms of accessing opportunities and equal pay within the labour market. Market reforms and cut backs on social welfare and child care have forced many women in Mongolia to take up low paying jobs in the informal sector, where the benefits of flexibility are traded off against wages and conditions. Child labour is also on the rise, particularly amongst those from poor, female-headed households. The ASM sector is one such sector witnessing an increase in both female and child labour (MEK Co. Ltd et al. 2005; MRPAM 2005).

---

1   Mongolia has witnessed an extraordinary increase in male mortality in the 20–34 age bracket, currently more than three times that of women in the same age bracket (NSOM 2007).

Gender and Employment: Formal and Informal Mining

According to the 2008 Mongolian country gender assessment (ADB Mongolia 2008), women employed in the formal mining sector receive on average 80.9 per cent of male salaries. In the informal sector, women receive considerably less—just 50 per cent of men's wages. However, women's ability to access work in the formal mining sector is considerably more difficult than it is within the ASM sector. Mining company job advertisements openly express a preference for male employees due to the general harshness of conditions, which are not thought to be suitable for women.

While jobs in mining and construction are high paying, the lack of enforcement of labour standards and safety measures has resulted in high levels of workplace deaths and accidents amongst employees, increasing the ranks of men in the long-term unemployed due to disability. This has increased the number of female-headed households and women's economic burden, compelling both women and children to accept exploitative and high risk forms of employment, such as ASM and semi-voluntary prostitution. Increased domestic and gender-based violence in Mongolia has also been associated with men's loss of economic power as well alarming rates of alcoholism.

In contrast to the large-scale mining sector, both men and women work in ASM in Mongolia – primarily from poor, rural areas. A socio-economic survey of artisanal miners in gold mining conducted by UNFPA (2007) found that overall 65 per cent of artisanal miners are men and 35 per cent are women. Within the sector men outnumber women by 71 per cent to 29 per cent in hard rock gold mining, while in placer gold mining men represent 61 per cent of miners and women 39 per cent. Placer gold mining is more accessible for women, as it has lower initial costs than hard rock mining and does not require working in groups of 4–5 people—placer gold miners can work alone or in smaller groups of two-three people.

The majority of artisanal miners (66.7 per cent) are involved in alluvial mining, with the remaining 33.3 per cent involved in hard rock mining. The age of informal miners ranges from 15–55, divided evenly with 50 per cent of the workforce below and 50 per cent above 33 years of age. Single, uneducated men make up a large proportion of those in the 20–25 age bracket. The majority of women (80.6 per cent) engaged in ASM are in the 35–39 age bracket. Of these women, 74.2 per cent have families. Men predominate in placer gold mining (61.3 per cent as compared with 38.7 per cent women) and are also engaged in hard rock mining in greater numbers as compared to women—71.4 per cent as opposed to 28.6 per cent. In terms of educational attainment, 34.6 per cent of artisanal miners have completed secondary education. More women than men have a complete secondary education—39 per cent for women as compared with 32.2 per cent for men—suggesting that for some women, working in ASM reflects difficulties in accessing skilled employment opportunities in the formal sector.

Women and men do equal work across a wide range of ASM activities, including ore sack transportation, gold separation using water or blowing, digging, lifting ore sacks from up to 20 metre deep holes, working inside deep holes and vertical tunnels, milling, crushing and sluicing. Accordingly, both men's and women's health are equally affected by the harsh weather conditions and low occupational safety of ASM work. Although high numbers of women are involved in mining and carry out the same tasks as men, they are still also largely responsible for reproductive work in the home, including food preparation, fetching water, gathering wood fuel, and caring for children, the elderly and sick. Women are also primarily responsible for running small shops, all of which creates a significant paid and unpaid work burden for women engaged in ASM.

The main gender issues in ASM in Mongolia are thus similar to those faced by women in other sectors—namely women's double work burden and lower levels of pay for the same work. Child labour is a problem for both genders, as is poor working and harsh weather conditions and the resultant health impacts, however younger children and girls are paid in kind (mainly food) while older male children are paid in cash, revealing that gender disparities in pay begin early on in life. For all age groups, wages are lower for girls than for boys and furthermore, in the case of gold and fluorspar mining, the boys represent over 80 per cent of the total child workforce (MEK Co. Ltd et al. 2005). Child labour in ASM can be seen as a gendered issue in so far as it reflects the increasing rates of poor, female headed-households and male bread-winner disability and mortality.

Given the high numbers of women engaged in ASM, a gender mainstreaming approach to the reform and regulation of the sector is warranted, in order to affectively address gender issues. At a government level, a range of gender-sensitive health programs and policies are needed for the sector. This should involve men and women participating equally in capacity building training programs, ASM community associations and actions, support to develop gender sensitive ASM institutions, programs and training manuals. It also means ensuring that men and women have equal rights and access in terms of the opportunities provided by improvements to ASM technology, the availability of credit and health services in mining areas or nearby mining areas.

MRPAM and SDC: Sustainable Artisanal Mining Project

The Mineral Resources and Petroleum Authority of Mongolia (MRPAM) recently developed a sub-program for the development of ASM in Mongolia which will run until 2015. The sub-program was approved by the same Government Resolution which approved the temporary regulation of ASM. The regulation has just been implemented, but it is hoped that feedback from local government authorities and artisanal miners will be reflected in policy documents as well as future improvements and modifications to the resolutions. The greatest achievement of these resolutions is that they recognise ASM as a legitimate economic activity and oblige government agencies to coordinate their activities.

To help deliver the program of reform to ASM, MRPAM has partnered with the Swiss Agency for Development and Cooperation (SDC) to manage and deliver the Sustainable Artisanal Mining Project (see Grayson et al. 2004; Janzen et al. 2007; SDC 2010). Project staff have been working with both the government and ASM communities to develop an appropriate legal framework, identify strategies to build the capacity of ASM communities and deliver education and training. The main goal of the project is to contribute to the development of responsible artisanal mining in Mongolia, improve the well being of artisanal miners (men and women) in a cost effective and sustainable manner and thus contribute to rural development and poverty reduction. In order to achieve this goal the project has identified four key objectives:

- To improve the development and implementation of a transparent and straight-forward policy and regulatory framework for artisanal mining.
- To improve the formation and functioning of institutional structures and organisations within artisanal mining at all levels.
- To strengthen the capacity of artisanal mining communities to engage in profitable and responsible mining and extended business activities aiming to reduce poverty.
- To empower artisanal miners and other resource users to address and solve existing and potential ecological as well as social conflicts responsibly.

To achieve these objectives, a community development team is currently working with ASM communities to develop formal organisational structures which will draft and implement community action plans. To this end, project team members facilitate meetings with community members to help identify community needs and interests and form work groups based on specific needs. ASM communities involved in the project have formed technical, social, and organisational development groups, comprising both men and women miners.

The project recognises that a multi-faceted approach which addresses social, environmental, legal and economic issues, is ultimately needed to make ASM a more sustainable activity. The project seeks to make ASM more cost-effective and therefore more profitable through the introduction of efficient and environmentally friendly tools and technologies. Technology transfer by project staff has included the introduction of chemical free processes for mineral extraction and processing as well as safer tools and technology. This has been coupled with ongoing education and training in safe work practices and occupational health and safety. Another project goal is to promote local economic development through the development of sustainable entrepreneurial entities (SMEs), which will generate formal employment and allow for diversification of labour and products (such as gold washing services, adventure tourism, production of mining equipment or jewelry).

Helping ASM to become an accepted, responsible activity requires dialogue and partnerships with local government authorities. The project is facilitating a process whereby miners will register as citizens, agreeing to pay taxes and entering into contracts to mine specific areas which have a range of clauses that must be adhered to, from upholding occupational safety standards to the rehabilitation of mine sites. In return, miners are able to vote and are offered social and health insurance and access to other government services. The aim of these partnerships is also to help reduce a long history of conflict between small-scale miners, government authorities and large-scale mining operations. At a later stage, it is envisaged that the project will support the formation of civil society organisations at regional and national levels to coordinate artisanal mining activities and represent the ASM community at those levels.

Gender has been mainstreamed throughout project activities in recognition of the participation of women in ASM and the link between the sustainability of ASM and gender roles and relations in ASM communities. All community institutions and structures created by the project must be gender-sensitive, including ASM associations. In terms of gender balance, the proportional participation of women and men in capacity building training, in the implementation of community action plans and in project work with groups of artisanal miners has been mandated. Gender-sensitive training programs and manuals have also been developed to ensure equal access, rights, responsibilities and opportunities for men and women in ASM activities. Gender analysis (the systematic gathering and examination of sex disaggregated data to identify and understand inequities based on gender) has also been a key and ongoing part of the project.

Conclusion

Since the transition to a market economy, the mining sector has become a vital part of the Mongolian economy. Although the government has opened large parts of the country up to mineral exploration and extraction by multi-national companies, capital-intensive, large-scale mining has boosted GDP, but not generated jobs. As a result, despite a sustained increase in GDP over the last decade, high levels of poverty and unemployment remain. At the same time, ASM has continued to expand and provides a livelihood for an increasing number of Mongolia's rural poor.

If managed in such a way that the social, economic and environmental costs and benefits can be balanced, ASM can contribute to the sustainable development of rural communities. ASM can create financial capital that can be invested in a range of diverse enterprises, build human capital in rural areas by increasing levels of education and technical skills, and generate social capital through redirecting the wealth generated into local economic development, health and welfare.

The World Bank has stated that mining is compatible with sustainable development, able to be achieved through the combination of efficient use of economic capital, maintenance of ecosystem integrity and natural resource productivity with social equity and mobility, participation and empowerment. In order to achieve this model of mining-driven sustainable development, governments need to assist by promoting equitable, participative community development as a basis for economic development. This includes gender equity through gender mainstreaming community development initiatives.

Mongolia's example shows that the regulation and development of ASM can be achieved, but only in partnership with government agencies, large-scale mining companies and other stakeholders. Given the complex web of social, legal and environmental issues involved in ASM, national and local governments and international agencies should assume the lead role in addressing them in partnership with ASM communities. In recognising the historical involvement and high levels of participation of women in small-scale mining in Mongolia, the Sustainable Artisanal Mining Project can also serve as an example of how gender can be integrated into mining-led development initiatives from the outset, in order to achieve a sustainable future for ASM and in so doing, for the many millions of people that depend on ASM for their livelihood.

References:

ADB (Asian Development Bank) Mongolia, 2008. 'Country Gender Assessment.' Unpublished report to ADB Mongolia. Ulaanbaatar: ADB Mongolia.

Grayson, R., T. Delgertsoo, W. Murray, B. Tumenbayar, M. Batbayar, U Tuul, D. Bayarbat and C. Erdene-Baatar, (2004). 'The Rise of "Ninja" Phenomenon.' Special Issue 'The People's Gold Rush in Mongolia'. *World Placer Journal* 4. Viewed 20 January 2011 at: http://www.mine.mn/WPJ4_1_Gold_Rush_in_Mongolia.pdf

Janzen, J., M. Priester, B. Chinbat and V. Battsengel (eds), 2007. *Artisinal and Small-Scale Mining in Mongolia: The Global Perspective and Two Case Studies of Bornuur Sum/Tuv Aimag and Sharyn Gol Sum/Darkhan-Uul Aimag.* Research Papers 4. National University of Mongolia: Ulaanbaatar.

MEK Co. Ltd, ILO and IPEC (International Labour Organization and International Program for Elimination of Child Labour), 2005. Baseline survey on child and adult workers in informal gold and fluorspar mining. Population Teaching and Research Center. Beijing: ILO.

MRPAM (Mineral Resources and Petroleum Authority of Mongolia), 2005. Sustainable Artisanal Mining.' Internal project document. Ulaanbaatar: MRPAM

———, 2007. 'MRPAM Annual Report.' Artisanal Mining Division Report. Ulaanbaatar: MRPAM.

NDS (National Development Strategy), (2008). 'Millennium Development Goals Based Comprehensive National Development Strategy 2007.' Report from task force to develop comprehensive national development strategy. Ulaanbaatar: Government of Mongolia.

NSOM (National Statistics Office of Mongolia), 2007. 'National Statistical Year Book 2007.' Ulaanbaatar: Government of Mongolia.

SDC (Swiss Agency for Development Cooperation), 2009. 'Sustainable Artisanal Mining Project.' Viewed 20 January 2011 at: http://www.swiss-cooperation. admin.ch/mongolia//ressources/resource_en_184875.pdf

UNFPA (United Nations Population Fund), 2007. 'Socio-Economic Situation of Artisanal Miners in Mongolia.' Ulaanbaatar: UNFPA.

UNDP (United Nations Development Programme) Mongolia, 2008. 'Mongolia Human Development Report 2007.' Unpublished report to the UNDP. Ulaanbaatar: UNDP Mongolia.

12. Gender Mainstreaming in Asian Mining: A Development Perspective[1]

Kuntala Lahiri-Dutt and Gill Burke

Introduction

Gender is one of the defining features of mining in Asian countries. This is for several reasons: the traditional involvement of rural and poor women, often from indigenous and ethnic communities, in the mining workforce; the important roles women play in providing for the subsistence of families, leading to a greater burden of negative impacts of mining on poorer women; and lastly, the increasing feminisation of poverty and the informalisation of women's work throughout the Asian countries. The rising commodity prices are expected to result in an expansion of mining in Asia, leading to an intensification of conflicts against social and environmental injustices. Whilst these unresolved issues remain crucial to improving the livelihoods of innumerable people in Asian countries, the expansion may also be seen as providing an opportunity to tackle gender equities that new mining projects might introduce or exacerbate in local communities. This chapter argues that mining in the Asian context needs to be gender-sensitive for socially-just development outcomes. In doing so, this chapter builds a 'development case' for gender mainstreaming in Asian mining and indicates the possible directions such a mainstreaming process should take.

Mining and Gender in Asia

Asia has a long history of mining metals, mineral ores and gemstones.[2] The involvement of women as part of the workforce has been one of the characteristic features of Asian mining. Historical records, including oral histories, have revealed that women and men laboured together in mines. Although protective legislation prevents women's fuller participation, women still participate and contribute to mining economies throughout most of Asia today. Women's participation in the mines in the past was not necessarily restricted to artisanal

---

1   We are grateful to Ms Sophie Dowling for her invaluable research assistance in writing this chapter. Thanks also to Dr David J. Williams for extracting occupational data from 2001 Census Reports of India, and other relevant information.
2   We have excluded the oil and gas sectors of extractive industries from this discussion.

and informal above-ground mines or in shallow operations. Evidence exists in Japan and India that women formed an important part of the workforce in large underground mining projects. Understanding Asian mining begins with this key fact.

A popular notion is that mining is a masculine endeavour, impacting on, victimising and marginalising women as a whole in areas where new projects are introduced. The fact of women's participation in a range of work in and around mines in Asia demolishes this myth based on an imagined masculinity associated with mining, blurs the rigid boundaries of gender roles, and illuminates 'industrial work' and 'economic activities' with a gendered light. Once the 'mineworker' is presented as a gendered subject, the mines reposition themselves as gendered places, and 'Asian mining' becomes contextualised as operating within a non-European social and historical milieu that can bring new understanding to both gender and mining.

This chapter analyses the policy implications of gendered work in Asian mining in the context of recent trends in capital mobilisation in the mining sector in Asian countries. The renewed economic focus on Asian nations[3], has drawn in multinational capital to break 'new ground' and open new resource projects on a gigantic scale.[4] The decline of mineral reserves in the developed world has shifted mineral and metal production to Asia and its newly growing economies have also shifted market demand to this continent. Cross-capital flows within Asian countries have also helped trigger this enormous mining expansion. As rising Asian economies take up the important roles of both producers and consumers of minerals and metals, a significant transformation is underway in global mining.

The mining industry is not unaware of these trends. Several industry conferences and meetings have taken place in the last few years showcasing the 'Asian' nature of mining requirements, expertise and capital-flows, and a new industry magazine—'The Asian Miner'—has been spawned. At the same time, international processes and standards of practice have also been developed, including industry guidelines with regard to environmental and social practices. These derive from the International Council of Minerals and Metals (ICMM), a Global Compact funded by nine major multinational mining corporations and the IFC/World Bank. Civil society organisations, such as Oxfam Community Aid Abroad as well as national bodies such as Jaringan Advokasi Tambang (Mining Advokasi Network—JATAM) and mines, minerals and People (m,m&P) are also

---

3   The Mining, Minerals and Sustainable Development (MMSD) report of 2002, 'Breaking New Ground', devotes a significant effort to outlining desirable practice in terms of human rights and social-cultural issues around Asian countries, where major mining activities are expected to impact.

4   Examples can be found across Asia especially in extreme areas such as Tibet, the Gobi Desert in Mongolia and Chinese Turkistan.

active in Asia. These groups have pointed out the almost uniform failure of mining companies to bring socially just development outcomes to Asian mining regions. Research by these organisations has found that mining impacts more negatively on women than men, while men often benefit from the increased work opportunities mining brings.

Although the debates around mining and society in Asia have yet to be 'gendered' in the sense of moving beyond examining the gendered impacts of mining, a rich reservoir of academic literature exists which discusses gender identities and roles in the extractive sector. Relatively new concepts are entering Asian mining literature; for example Corporate Social Responsibility (CSR) is an increasingly-heard term. Thus there is a need to move beyond the 'impacts of mining on women' and to explore how mining could be equally just for both women and men in Asian societies. Gender inclusiveness and equality are integral parts of development. As a key development 'agent' in many less-developed countries and regions, can the mining industry avoid this debate, and if not, what should be its rationale? This chapter aims to explore this question in the general context of Asian mining.

This discussion traces women's agency in a most unusual place of work; just as Asia is an unexpected context for mining. Historically, a simplistic dualism has characterised mining literature; mining itself has been seen as the 'other' of 'normal' farming-based life. Similarly, miners in Europe have been characterised as the quintessential 'others' to 'normal' social life (Burke 1993). At times, this 'otherness' is manifested in class struggles between miners and other groups involving militant or collective behaviour, at others times, it has been manifested in the stereotype of the 'larrikin' mine worker, who has a tendency toward drunkenness, gambling and rioting. These dualistic images have also been highly gendered, pitching the 'hard, unrefined men, distinct and separate from other workers, hewing in mysterious dungeons' (Allen 1981) as the 'other' of women 'hewing cake' at home (Gibson 1992). Examining this 'otherness' of women in mining communities, Gibson (ibid.) found that women's own working class identity grew out of supporting their men. The gendered division of labour in mining communities also created a gendered spatial division between the home and workplace (McDowell and Massey, 1984). In this chapter, we contest this view and challenge the binary view of men=miner and women=home keeper by drawing attention to the fluid gender identities in Asian mining.

Can one talk about a distinctive 'Asian' mining today? Not much of the exploitative and/or subsistence-based 'Asiatic mode of production'[5] that Marx observed is valid in modern mining any more. Modern capital, as we noted, is moving to Asia not only from the more 'central' regions of Europe and North America but also across Asia from within. As a consequence, older conceptions relating to centre-periphery relations are no longer strictly valid in the field of mining. This raises the question of whether the modern capital-intensive interventions can justify the use of the term. The diversity of Asia's mining history, extraction characteristics and mineralogy is enormous. We begin with an acknowledgment of the heterogeneity of Asia in every aspect of its history, economy, society and culture. We accept that some of the 'Asian' features of mining in this diverse context would doubtless also be present in similar measure in the less developed parts of Africa and Latin America. Thus we use 'Asia' as a subset for particular part of the world, a convenient category.

Asia's mining history is of immense antiquity. For many centuries mining flourished throughout the continent, but was scattered and relatively small scale and artisanal in nature. Colonialism in the nineteenth and twentieth centuries transformed mining by introducing modern industrial operations. Although colonial mines did not represent an area of capital entry on the scale of agricultural plantations and industries such as jute, they were significant in bringing these peripheral countries into contact with the centre. British Malaya and Dutch Indonesia rapidly became the world's largest tin producers. Coal from French Vietnam formed an important part of global supply. Moreover, these early colonial-era mines began a pattern of mining-led development within Asia. Independent countries such as Siam followed the same mining development model, whilst Japanese coal production expanded rapidly after the Meiji Restoration.[6]

This pattern of mining sector industrialisation and production expansion continued well after the colonised countries became independent, although by the late twentieth century there were some notable changes in the overall mining profile. Tin mining in Malaysia and coal mining in Japan have both ceased and coal mining is now more significant than tin mining in Indonesia. Yet, farming has remained the predominant occupation and only a few Asian countries (as opposed to African countries) can actually be termed 'mineral

---

5   The term 'Asiatic Mode of Production' was originally coined by Marx to account for a type of society outside the mainstream of Western development, as the term 'Asiatic' was not restricted to the geographical area of Asia. However, the concept came to be one of the most controversial because of the implication that the 'Asiatic' (or feudal) societies could not reach socialism without going through the purgatory of a capitalist development process (see Bailey and Llobera, 1981, for a detailed outline of the debates around the concept).
6   In 1874 Japanese coal production was 280 000 tonnes. By 1919 this had risen to 31 million tonnes (Hane 1982: 227).

dependent'[7] in terms of scale of mineral production or values. The ex-Soviet states in Asia—especially Kazakhstan, Tajikistan and Mongolia (MMSD 2002: 45–6), figure more highly in ore and metal export dependence than most others.

As stated, historically, mining in Asia has been characterised by the involvement of women. Feminist historians such as Sone (2006: 154–5) and Hane (1982: 233–6) have both shown that women formed a major part of the colliery workforce in Japan until around 1946 when they were completely banned. In colonial India the 'modern' coal mining economy also depended heavily on women's labour, often as part of family labour units (Lahiri-Dutt 2000). In more recent times, modern mechanised mines have hired women to operate trucks and other heavy equipment. In some Indonesian coal and gold mines, women operators drive state-of-the-art, mechanised shovels and dozers in open cut mines. At the same time, prohibitions against women workers in industrialised mining gradually relegated women to artisanal and small-scale mining (ASM) and the informal sector (Lahiri-Dutt 2007). Although such forms of 'peasant mining' have a long history in Asia, in many countries ASM is not always officially recognised.[8]

Mining Countries, Gender Equity and Development

With its long history of gemstone extraction and other forms of mining, China is by far the largest player in industrial mining in Asia, if not globally. In terms of overall scale and magnitude, China's mining industry ranks third in the world, although production statistics remain sketchy[9]. In terms of domestic market-oriented mineral production, women are under-represented in the mining sector in China as compared to other sectors—at least according to official statistics. Estimates are that women comprise only around three per cent of mining and quarrying workers in urban areas (Yao 2006: 239). A 1936 law prohibited women from working underground in mines, although they can still be deployed at the surface and appear to hold a reasonable number of administrative and technical

---

7 Mineral dependence is the extent of reliance on mineral output as a proportion of gross domestic product (GDP) or the value of minerals in relation to exports and having a close relation to a countries' general level of economic development.

8 In Indonesia for example, small-scale mining and quarrying for industrial minerals by family groups was recognised through the 'People's Mining' Act (Pertambangan Rakyat), but mining for precious metals is still illegal. Consequently, there still are some legal artisanal diamond mines, but ASM tin is illegal.

9 China has about 80 000 state-owned mining enterprises and 200 000 collectively-owned mines. China is now the largest coal producer in the world, extracting almost 2500 metric tonnes in 2006 (WCI 2007). However, more than 90 per cent of this production came from 22 000 small mines (less than 300 000 tonnes per year) and more than half from 12 000 very small mines. Although it exports a small amount of coal mainly to South Korea and Japan, the rapid expansion of the country's economy has made China a net importer of coal (38 metric tonnes in 2006) which has led it to impose restrictions on coal exports.

positions—as high as 35 per cent. However, Yao has found there is continued illegal employment of women underground in a range of private and state owned mining operations in China (ibid.: 249–50).

Although India cannot be described as a 'mining country'[10], the gamut of mineral reserves there, the large number of people who earn a livelihood from mining, and the importance of minerals in sustaining economic growth has made mining a key sector. India is also one of the leading mineral producers in the world; it is the largest producer of mica, second largest producer of chromites, third largest producer of coal and lignite, fifth largest producer of iron ore, and sixth largest producer of bauxite and manganese ore. In recent years, the share of the Mining and Quarrying sector (as referred to by the Census of India, 2001) in India's gross domestic product (GDP) has grown at an average rate of five per cent, and this rate is likely to continue to grow.

Currently, women comprise around 14 per cent of workers in the Indian mining and quarrying sector (ibid.). However, the extent of informalisation of women's labour is evident from the much higher proportion—33 per cent of all workers—amongst those defined as 'marginal workers'.[11] In State-owned coal mining, women comprise around 5.6 per cent of the workers, but around 17 per cent of marginal workers in the same sector are women. The participation of women is highest in dolomite mining (33 per cent), mica mining (25 per cent), clay mining (23 per cent), stone quarrying (23 per cent), salt extraction (23 per cent), manganese ore mining (21 per cent) and gem stones mining (19 per cent), indicating that women's labour is concentrated in the small-scale or informal mining sector. In gold mining, women comprise 57 per cent of marginal workers, with chromium ore mining also employing women as 38 per cent of the marginal workers. However, official data on mining employment may not reveal the full extent of women's participation in mining activities in India or in other Asian countries. The construction boom that has accompanied Asian economic growth has massively increased demand for building materials such as stones, sand and gravel, which are mostly obtained from small mines and quarries. Since women's labour is concentrated in the informal sector, one can assume concomitant increases in their employment.

Other data on women's employment trends in Asia reveals that gender inequality is particularly concentrated in the mining sector. For example, in

---

10   Such definitions are restricted to those countries whose wealth is largely derived from the extraction of mineral resources. Australia, Canada and PNG are amongst such countries. Only around three per cent of India's GDP comes from mining and minerals.

11   According to the Census of India (2001), 'marginal workers' were those who did not work for a major part of the year preceding the enumeration (less than 183 days or six months). Marginal workers are often landless labourers or farmers engaged in various informal jobs during the non-cropping season 'Main workers' were those who worked for a major part of the year preceding enumeration, that is those engaged in any economically productive activity for 183 days (six months) or more during the year.

the Philippines, the only Asian country to rank well in the Global Gender Gap Index, the number of registered women mining engineers from 1927–2003 was 40. The number of registered male mining engineers for the same period was 2 639 (Chaloping-March 2006: 189). The different work performed by men and women and the number of both employed in the mining sector differs significantly to overall national participation rates. According the Global Gender Gap report (WEF 2008), although Philippine women outnumber men as legislators, senior officials, managers, professionals and technical workers, in mining they are concentrated in administrative work—where their numbers are still less than men. Two mines surveyed by Chaloping-March (2006: 193) in 2000 revealed that of 152 managerial positions, only 16 were held by women. Within the production area in the same mines, there were 3 714 men employed and only 51 women. This might be explained by the perception in the Philippines as elsewhere, that underground and earth work is out of bounds to women, but does not explain women's unequal employment in the administrative and technical areas of mine operations.

It is well-known that a low level of gender equity is generally related to low levels of economic development. In the mining literature, we also find that high numbers of women engaging in ASM and informal mining is often an indicator of low levels of gender equity. In less developed countries, the labour force is increasingly becoming feminised, but at the same time, women are being employed in more casual jobs. In their study on Africa, Hinton et al. (2006: 13) observed that the gender roles and the status of women in ASM reflected broader societal levels of gender equity and women's empowerment, including: 'women's and men's access to and control of, resources; their ability to attain knowledge or resources, their decision-making capacity or political power; and beliefs or attitudes that support or impede the transformation of gender roles.'

In Asia women also participate in ASM in greater numbers than in formal mining, which suggests that the correlation between gender roles in ASM and those found in the wider society may be a global phenomenon. Around the world the proportion of women engaged in informal mining varies between 10–40 per cent (Lahiri-Dutt, 2008). Levels of participation vary according to the nature of the mineral, production technology, the physical location of the operation and above all the length of time women have been involved in mining. Although it is difficult to find accurate data on the levels of participation and production in informal small mines and quarries, the number of women engaging in ASM throughout Asia is thought to be rising (Caballero 2006; MMSD 2002: 21).

To further explore the correspondence between gender equity and development in Asia, we now turn our attention to some quantitative and qualitative indicators. Simple national performance statistics such as the Global Gender Gap can reveal the close correlation between a countries' level of economic and social development and gender equality. Of the participating Asian countries in the 2007 World Economic Forum (WEF) survey[12], the Philippines ranked sixth and outshone many developed countries, such as the Netherlands (twelfth) and Australia (seventeenth). However the gender equality rankings of other participating Asian countries were generally low. Those with large mining sectors were particularly so—for example China ranked 73, Indonesia 81 and India 114.[13] A comparison of the WEF rankings with the UNDP's Human Development Index (HDI)[14] demonstrates the positive correlation between overall levels of human development and gender equality. The UNDP Gender-Related Development Index[15] (GDI) highlights the differences between development outcomes for men and women. In terms of individual Asian countries, both the UN's 2007/2008 GDI report and the Global Gender Gap Report rank China 73. Women's labour force participation is 76 per cent as compared with 88 per cent for men but women's income is estimated to be just over one half of men's. India's UN GDI rank is 114, closer to the Global Gender Gap Report rank of 113. Similarly, opportunity and income levels show considerably high levels of inequality as the average income of women is just over one quarter that of men's. The female labour participation ranking for India, low as it is (36 per cent for women as compared with 84 per cent for men), does not reveal the full extent of female participation since it does not include marginal workers or those in the informal sector. Indonesia ranks 81 in the Global Gender Gap Report, but 94 according to the UN's GDI. Indonesian women's labour force participation is 53 per cent to men's 87 per cent and their income is estimated to be on average just less than one half of men's.

---

12   WEF annually surveys gender differences in attainment in four key areas (economic participation and opportunity, political empowerment, educational attainment and health and survival) between men and women across 128 countries in the world. Of the ten highest ranked countries, eight are from west and northern Europe. Comparisons between the 2008 Global Gender Gap Index and the Global Competitiveness Index and GDP per capita confirm this correlation. The report's authors conclude, 'while correlation does not prove causality, it is consistent with the theory and mounting evidence that empowering women means a more efficient use of a nation's human talent' (WEF 2008: 20) .

13   Disaggregated scores for the four separate areas measured reveal differences hidden by the aggregate country rankings. China for instance is ranked 60 for women's level of economic participation and opportunity, but is actually close to the bottom in terms of health and survival at 124. India has a particularly poor rank for economic participation and opportunity at 122—close to the bottom of the scale—but ranks 21 for women's political participation, a result of the Constitutional provision reserving 33 per cent of seats for women in local level elections. Only the Philippines retains a consistently high rank in terms of outcomes for women across all four areas. Among regions of the world, Asia scores second lowest in terms of women's economic participation and opportunity—behind Latin America and Sub-Saharan Africa. Performance in terms of women's educational attainment and health and survival is similarly low relative to other global regions.

14   The Human Development Index or HDI is a composite index that measures a country's achievements in three basic areas of human development: health, knowledge, and a decent standard of living. Health is measured by life expectancy at birth; knowledge is measured by a combination of the adult literacy rate and the combined primary, secondary, and tertiary gross enrolment ratio; and standard of living by GDP per capita (Purchasing Power Parity US$).

15   Each country's GDI is worked out as a proportion of their HDI. The GDI rankings are similar to the Global Gender Gap Index and the HDI—with predominantly the same developed countries making up the top 20, with some variations in place of rank.

Qualitative differences in terms of variables help to explain Indonesia's different rankings. The higher Gender Gap Report ranking is due in part to the relatively higher number of Indonesian women that are found in professional and technical work—42 per cent as compared with 58 per cent of men. In India only 21 per cent of women as compared with 79 per cent of men are employed within this work sector. In Mongolia, this variable also explains a significant difference in rankings, with the UN GDI ranking Mongolia 100 and the Global Gender Gap Report ranking it 62. Mongolia demonstrates some surprising gender trends in terms of economic participation and opportunity. The level of female labour force participation is 56 per cent to men's 83 per cent, literacy levels are equal between women and men at 98 per cent. A higher proportion of Mongolian women are employed in professional and technical work—66 per cent as compared with 34 per cent of men, however women make up only 30 per cent of legislators, managers and senior officials compared with 70 per cent of men, and—despite impressive levels of equality in terms of education and literacy—women's average earnings are half that of men.

Mongolia's example brings us to qualitative variables that are not included in quantitative ranking exercises to ascertain gender equality (save for the focus on income) These include the often stark but invisible differences between men and women in accessing income, resources and services—the so-called 'feminisation of poverty'.[16] Understanding why women dominate the ranks of the world's poor requires analysis of poverty as a social and cultural phenomenon—not simply an economic one. Early feminists recognised that women's lower status reflected a wide range of impediments and challenges—not least the challenges of bearing and raising children and the responsibilities of managing a household—particularly following a husband's death, divorce or abandonment.

Gender Mainstreaming in Asian Mining

The mining industry has generally stayed outside of the gender debate, although recent efforts highlighting women's productive roles in mining may help bring about some change to this. From 2000 onwards international conferences organised by academics, international agencies and NGOs have explored ways to include mining in gender debates.[17] They also clearly agreed on the importance

---

16   Feminisation of poverty was one of the key issues debated in the Beijing Platform for Action and was a focus when preparing the Millennium Development Goals.

17   The Australian National University funded a 2004 international workshop on Women Miners in Developing Countries. The International Women and Mining Network (RIMM) initiated by grassroots co-operative mineworkers in Latin America, organised three successive and widely attended conferences in the Philippines, Bolivia and Vizag. Oxfam Australia's International Conference 2002 resulted in an edited publication, 'Tunnel Vision: Women, Mining and Communities'. Two successive World Bank conferences in 2003 and 2005 were held in Papua New Guinea (PNG).

of bringing gender issues into the purview of the global mining industry. Two recent books show that contrary to the popular view of mining as an exclusively male endeavor, women have contributed and still contribute significantly to this industry (Lahiri-Dutt and Macintyre 2006, Gier and Mercier 2006). However, these efforts have yet to result in discernible changes to the industry's policies, practice and outlook. Gender continues to remain at the periphery of the highly corporatised world of mining as evident from the complete lack of attention to gender in the International Council for Minerals and Metals (ICMM) 2005 'Community Development Toolkit'. This begs the question why Asian countries should undertake gender mainstreaming measures and how they should go about it.

At this point the global consensus for the need to mainstream gender in any development process needs considering—particular so in the Asian context. The relationship of gender to development has continued to grow in significance over the years, along with international debate about the role and status of women and their contribution to economic development. Since 1975, which marked the International Year of Women, the focus has shifted from 'women and development' to 'gender and development' (McIlwaine and Datta 2003). This reflects a number of important ideas that have emerged over this period— the importance of particular historical, social and cultural trends shaping what we call gender, the roles of men and women in a given society, and the recognition that both women and men contribute to the transformation of their respective status, roles and opportunities. The recognition that social and cultural institutions, including government and religion, play important roles in defining gender roles and relationships has also shaped new approaches to the achievement of gender equity.

Gender mainstreaming as a concept is not new; it was first proposed at the 1985 World Conference on Women in Nairobi and more formally developed at the 1995 World Conference on Women in Beijing. In practice it is a strategy for assessing and making women's as well as men's concerns and experiences an integral part of any institution or operation. The ultimate goal is to achieve gender equality. According to Walby (2005: 321), gender mainstreaming is a practical strategy involving the reinvention, restructuring and re-branding of a key part of feminism, comprising, 'both a new form of gendered politics and policy practice and a new gendered strategy for theory development'. It invokes a vision of gender equality achieved through the mainstreaming process by drawing on notions of 'sameness', 'differences' and 'transformation'.

Since 1995 gender mainstreaming has been established as an intergovernmental mandate, re-affirmed in the follow-up UN General Assembly special session held in June 2000. Almost all member states have begun adopting gender mainstreaming in consensus as a global strategy for promoting gender equity and implementing gender mainstreaming across a wide range of national policy areas and institutions. Most Asian countries have now initiated policies and programs designed for implementation in a systematic way, with an embedded approach to gender. Similarly, gender mainstreaming strategies have been adopted by all the international aid agencies—the United Nations, World Bank and International Labour Organisation, and also by the Council of Europe. Practitioners in several development sectors, such as water management, have undertaken gender mainstreaming efforts. The process has not been trouble-free, and even now both the concept and its practice remain contested (Daly 2005). For example, the prioritisation of Millenium Development Goals in less developed countries can mean that the competitiveness of an economy takes precedence over equality considerations, thereby perpetuating women's ranks amongst the lower paid in many parts of the developing world.

Can one talk of gender mainstreaming in the mining industry? Mining clearly is a 'gendered industry'. Women are already extensively represented in both the formal and informal mining sectors from the highest to the lowest level. The fact that a woman took up the position as CEO of Anglo American plc, the world's third largest mining multinational, might be seen as evidence that the promotional glass ceiling had been well and truly smashed.[18] Women in developed countries train as geologists and mining engineers and, in some instances, work as skilled miners. This is less common in Asian countries but gradually many urban and educated women are coming to occupy skilled and even high-level managerial positions within the industry.

However significant, these incremental achievements fall well short of the requirements for gender mainstreaming, which takes a comprehensive approach to gender equity, and requires embedding gender-sensitive practices and policies throughout an organisation's internal structures and processes, as well as external activities. At the upper levels of the formal mining sector women can be said to have gained their position by performing to standards set by men (Rossilli 1997). This is often the case from the beginning of their careers. Although women are often actively recruited to compensate for falling male enrolments, the course syllabus at mining and engineering schools are rarely gender sensitive. As far as can be ascertained, no mining company or corporation has yet introduced gender mainstreaming policies. It is clear that a transformation is required amongst the major players in Asian mining. The key question is whether this can be achieved from within the industry or whether it must wait for further changes in the wider society.

In terms of gender mainstreaming measures in corporatised mining, women could become 'incorporated' in the bottom layer through establishing a 'quota'

---

18   Although later developments proved that sexism did not end within that company with her appointment.

for women and eventually some jobs would come to be designated as 'women's jobs'.[19] To avoid the pitfalls of developing initiatives that are little more than tokenistic exercises in gender equity, it is necessary to get the philosophy right with the correct nomenclature established from the outset. It also means that targeted gender initiatives to address areas of inequality and specific priorities for women and men must be taken up. Women-targeted initiatives can be used to complement gender mainstreaming to ensure program sustainability, but must also be supported by a comprehensive institutional framework approach to gender equity.

In Asian mining, gender mainstreaming should involve the assessment of the entire gamut of mining activities as gendered practices, including corporatised, artisanal and informal mining. It should also involve investigating the potential gender impacts of any new mining activity and the integration of men's and women's needs and concerns into any proposed course of action, such as community development projects, procurement chains or mine closure plans. The establishment of gender-sensitive mining and development policies and initiatives by both government and industry are necessary preconditions for achieving these goals.

Approaches to Gender Mainstreaming in Mining

The current philosophy and approach to gender taken by the extractive industries can be said to reflect three broad approaches to gender. These can be described as the development approach, the human rights approach and the efficiency approach or 'business case'. The development approach is taken by Mason and King (2001) to mean gender equality is a primary goal in all areas of social and economic development. It is a core development issue, 'a development objective in its own right' and 'essential for reasons of fairness and social justice'. Evidence of the importance of this can be found in the ASM sector throughout the world (see the growing number of publications on gender and informal mining practices by Hinton et al. 2003, 2006; Yakovleva 2007; and Lahiri-Dutt 2008). A large number of researchers and those engaged with informality have arrived at the conclusion that gender must be recognised and integrated into broader development policies and initiatives in ASM. The large scale mining industry, however, continues to see 'the community' as ungendered and homogeneous, and remains ambivalent about women's participation in mining.

---

19   Rees (1998) found that this was largely the way that gender mainstreaming operated in the European Union.

The evidence base for gender mainstreaming pursued by the major international bodies such as the World Bank's Oil, Gas and Minerals Division derives from the now large amount of research and literature detailing the 'impacts of mining on women'. These studies show that there is a gender bias in the distribution of risks and benefits from mining projects: benefits accrue to men in the form of employment and compensation, while the costs fall disproportionately heavily on women. Similar findings run throughout other studies done in Asian countries inviting foreign mining capital to operate around subsistence-based communities. In her early ethnographic work on the political economy of development in a mining town in Indonesia, Robinson (1986: 64) noted these changes were gendered, that 'in the change from peasant agriculture to wage labour the women have been subject to a decline in their economic independence'. At the same time, Robinson found, 'women have become more economically dependent on men, changes in cultural forms of the expression of gender have resulted in a decline in some of the restrictions on women's personal freedom which hitherto prevailed in the community' (ibid.).

The efficiency approach or 'business case' for gender equity in mining proposes it as 'good for business', economic growth, and poverty reduction. This approach is rooted in the largely unsubstantiated argument that women are 'better workers' than men.[20] The World Bank's 2007 Gender Action Plan also follows this line; placing emphasis on gender equality as 'smart economics' and notes that it helps to remove gender-based discrimination to get more efficient business outcomes, because gender discrimination prevents both women and men from reaching their full potential. It is true that the employment of women can do more to encourage global growth than increases in capital investment and productivity improvements (see for example AusAID 2006), but it does not necessarily make women 'better workers' than men in mines even though the Minerals Council of Australia's 2007 publication 'Unearthing New Resources' represents women as the new resource for skills-hungry mining companies.

The rights-based approach to gender mainstreaming in mining is exemplified in the work of Macdonald (2006) and elsewhere in the work of Oxfam's Mining Ombudsman, a range of other mining-focused NGOs and international donor agencies. This approach derives from the Universal Declaration of Human Rights, which states: '(t)he human rights of women and the girl-child are an inalienable, integral and indivisible part of universal human rights'. In India, the rights-based approach to gender mainstreaming in mining has not been fully developed, NGOs and their associates being more concerned with social

---

20   A whole range of such assumptions, often captured in glossy texts with catchy headings, have been compiled by Harcourt (2006: 13–4) who has critiqued claims such as: 'train poor women for a job and your investment is guaranteed', 'replace poor women with men in a factory and you have a docile and effective workforce for less pay and less trouble'. Such claims are dangerous because they create an image of all women as 'special workers' with special abilities that antagonise men.

and environmental justice issues. Within civil society organisations, inner tensions remain which can hamper efforts to bring attention to and address gender equity in relation to mining. These tensions within the 'women and mining' movement reveal a need to move away from essentialising women as 'victims' to seeing women as economic agents, bound within differences of class and ethnicity.

Conclusion

We have shown that despite common perceptions, mining is not a 'non-traditional' area of work for women in many parts of Asia. Women's work in Asian mines also raises the importance of understanding the relationship of class and caste to gender, as mining work is seldom undertaken by educated urban women. That there is no universal category of 'women' in Asia and that 'work' is a defining feature of life for women are essential to understanding gender in Asian mining. It is thus important to bring ethnicity and social realities into focus to properly to debate how gender can be mainstreamed in the extractive industries. Unfortunately many women continue to have the least voice, weakest bargaining power, and are often invisible even to the staunchest feminists. For example, the work demands of women working in metro-based Information and Communication Technology (ICT) have inspired changes in Indian labour laws[21], but at the same time, the last Labour Commission observed that women must be denied underground mining work (Lahiri-Dutt forthcoming). Women workers in many informal mines work within family labour units, thus it is important to recognise this work forms part of household livelihood strategies on which several family members depend, and ensure that women's access to mining work is supported and not undermined. Because 'women and children' in many Asian countries are legislatively clubbed together, efforts such as ILO-IPEC's Child Labour eradication projects have the potential to hinder women's rights to better conditions at work.

In 'engendering' the Asian extractive industries sector we need to remember women's long history of involvement in mining work, as well as tackling the challenges of the gendered impacts of mining and incorporating a gendered

---

21   The 2002 Report of the National Commission of Labour in India notes in its several recommendations in point 6.121 of the Review of Law section (NCL, 2003: 95–6): 'We would recommend enactment of a general law relating to hours of work, leave and working conditions, at the workplace.' However, this 'omnibus law', according to the Commission, 'should' incorporate the 'prohibition of underground work in mines for women workers, (and) prohibition of work by women workers between certain hours'. This well-intentioned report completely ignores women workers in mining although urban-based working-class women such as rag-pickers, construction workers and *bidi* (indigenous type of cigarette) workers do get their due consideration. It does consider the question of night work by women at some length, although the focus of this attention is clearly on the information and communications technology workers.

outlook in community outreach projects undertaken by mining companies.[22] It is also important to remember the generally restrictive environment within which women work in Asia. As shown in this chapter, in many Asian societies, women still cannot own, hold or transfer assets, apply for credit and are not recognized or able to lay claim to the benefits of a parents' will, or work freely without legal restriction or without the permission of male relatives. Consequently, women's status and economic wellbeing were and in many instances still are entirely dependent on their husband, father or other male relatives. Women's lack of legal, social and political rights all impact on their status, their ability to participate freely and fully in society, to exercise choice and agency and to increase their economic productivity and wealth. At the same time, the number of women engaged in the informal mining sector—characterised by greater insecurity and lower wages—has increased, putting them at greater risk of poverty. In this context, does the extractive industry continue to see women as victims or help them to flourish as full economic citizens, empowering them to build families and communities that are economically strong and sustainable on a long term basis? We hope this chapter will form the basis for a more informed debate on gender mainstreaming in the production and in the outreach processes of the extractive industries in Asia.

References

Allen, V.L., 1981. *The Militancy of British Miners.* Shipley: The Moor Press.

AusAID (Australian Government Overseas Aid Program), 2006. 'Gender Equality in Australia's Aid Program: Why and How.' Canberra: AusAID. Viewed 15 January 2011 at: www.*ausaid*.gov.au/publications/pdf/*gender_*policy.pdf

Bailey, A.M. and J.R. Llobera (eds), 1981. *The Asiatic Mode of Production: Science and Politics.* London: Routledge.

Burke, G., 1993. 'Asian Women Miners: Recovering Some History and Unpacking Some Myths.' Paper presented to the *Women in Asia Conference*, University of Melbourne, 1–3 October.

Caballero, E., 2006. 'Traditional Small-scale Miners: Women Miners of the Philippines.' In K. Lahiri-Dutt and M. Macintyre (eds), op. cit.

---

22 The concept of 'community development' has not quite matured in the extractive industries sector excepting a handful of (usually) privately owned companies taking up development projects in surrounding communities. That does not necessarily mean that the public sector companies do not have social awareness at all; many of these responsibilities have been ingrained either in precedence or tradition although only a few are written into formal laws. For example, jobs have traditionally been offered to 'land losers' although the number of these jobs have been declining and the awareness that besides those losing land (that is directly affected), there are a large number of people who are 'project affected'.

CASM (Communities and Small-Scale Mining), 2004. *Women and Children Workshop*. Communities and Small Scale Mining Annual General Meeting in October, 2004. Colombo: Sri Lanka.

Chaloping-March, M., 2006. 'The Place of Women in Mining in the Cordillera region, Philippines.' In K. Lahiri-Dutt and M. Macintyre (eds), op. cit.

Census of India, 2001. 'Final Population Totals, Government of India, New Delhi.' Viewed 20 January 2011 at: http://www.censusindia.net/

Daly, M., 2005. *Gender Mainstreaming in Theory and Practice*. Oxford: Oxford University Press.

Gibson, K., 1992. 'Hewers of Cake and Drawers of Tea: Women, Industrial Restructuring and Class Processes on the Coalfields of Central Queensland.' *Rethinking Marxism* 5(4): 29–56.

Gier, J. and L. Mercier (eds), 2006. *Mining Women: Gender in the Development of a Global Industry, 1670–2005*. New York: Palgrave Macmillan.

Hane, M., 1982. *Peasants, Rebels and Outcasts: The Underside of Modern Japan*. New York: Pantheon Books.

Harcourt, W. 2006. 'Making Change Happen' Editorial, Women's Rights and Development, *Development,* 49(1): 1–6.

Hinton, J., M.M. Veiga and C. Beinhoff, 2003. 'Women and Artisanal Mining: Gender Roles and the Road Ahead.' In G. Hilson (ed.), *The Socio-economic Impacts of Artisanal and Small Scale Mining in Developing Countries*. Netherlands: Swets Publishers.

———, 2006. 'Women in Artisanal and Small-Scale Mining in Africa.' In K. Lahiri-Dutt and M. Macintyre (eds), op. cit.

Lahiri-Dutt K., 2000. 'From Gin Girls to Scavengers: Women in Raniganj collieries, *Economic and Political Weekly* XXXVI(44): 4213–21.

———, 2003. 'Not a Small Job: Stone Quarrying and Women Workers in the Rajmahal Traps in Eastern India.' In G. Hilson (ed.), *The Socio-economic Impacts of Artisanal and Small Scale Mining in Developing Countries*. Netherlands: Swets Publishers.

———, 2006. 'Mining Gender at Work in the Indian Collieries: Identity Construction by Kamins.' In K. Lahiri-Dutt and M. Macintyre (eds), op. cit.

————, 2007. 'Role and Status of Women in Extractive Industries in India: Making a Place for a Gender Sensitive Mining Development'. Social Change, December, 37(4): 37-64.

————, 2008. 'Digging to survive: Women's livelihoods in South Asia's small mines and quarries.' *South Asian Survey* 15(2): 217–244.

————, (forthcoming). 'Gender and Labour at the Coalface: Making a Place for Women in the Indian Collieries.' In S. Raju (ed.) *Feminist Studies in Geography*, Oxford University Press: New Delhi.

Lahiri-Dutt, K. and M. Macintyre, 2006. 'Introduction.' In K. Lahiri-Dutt and M. Macintyre (eds), *Women Miners in Developing Countries: Pit Women and Others*. Aldershot: Ashgate.

McDowell, L. and D. Massey, 1984. 'A Woman's Place.' In J. Allen and D. Massey (eds), op.cit.

McWilwane, C. and K. Datta 2003. 'From Feminising to Engendering Development', *Gender*, Place and Culture, 10(4), *2003*, 369–82.

Mason, E.M. and A.D. King, 2001. *Engendering Development through Gender Equality in Rights, Resources, and Voice*. Washington D.C.: World Bank.

Massey and J. Allen (eds), 1984. *Geography Matters: A Reader*. Cambridge: Cambridge University Press/Open University.

MMSD (Mining, Minerals and Sustainable Development), 2002. *Breaking New Ground*. London: IIED and Earthscan.

NCL (National Commission on Labour), 2003. *Reports of the National Commission on Labour, 2002–1991–1967*. New Delhi: EconomicaIndia.

Parthasarathy, M., 2004. 'Exploring the Impact of Chromite Mining on Women Mineworkers and the Women of Sukinda Valley, Orissa: A Narrative.' In S. Krishna (ed.) *Livelihood and Gender: Equity in Community Resource Management*. New Delhi: Sage Publications.

Rees T., 1998. *Mainstreaming Equality in the European Union, Education, Training and Labour Market Policies*. London: Routledge.

Robinson, K. 1986. *Stepchildren of Progress: The Political Economy of Development in an Indonesian Mining Town*. Albany N.Y.: State University of New York Press.

Rossilli, M., 1997. 'The European Community's Policy on the Equality of Women: From the Treaty of Rome to the Present.' European Journal of Women's Studies 4(1):63-82.

Sone, S., 2006. 'Coal Mining Women Speak Out: Economic Change and Women Miners of Chikuho, Japan.' In J. Gier and L. Mercier (eds), op. cit.

Walby, S., 2005. Gender mainstreaming: Productive Tensions in Theory and Practice.' Social Politics 12(3): 321–43.

WEF (World Economic Forum), 2008. 'Global Gender Gap Report.' Geneva: World Economic Forum.

Yakovleva, N., 2007. 'Perspectives on Female Participation in Artisanal and Small-scale Mining: A Case Study of Birim North District of Ghana', Resources Policy 32(1/2): 29–41.

Yao, L. 2006. 'Women in the Mining Industry of Contemporary China', in K. Lahiri-Dutt and M. Macintyre (eds), op.cit.

www.ingramcontent.com/pod-product-compliance
Lightning Source LLC
Chambersburg PA
CBHW061244270326
41928CB00041B/3409

* 9 7 8 1 9 2 1 8 6 2 1 6 8 *